DESIGNING AND USING
MATHEMATICAL TASKS

DESIGNING AND USING MATHEMATICAL TASKS

John Mason, Professor of Mathematics Education at
The Open University

Sue Johnston-Wilder, Senior Lecturer in Mathematics
Education at The Open University

tarquin publications

The Open
University

This publication forms part of an Open University course ME825 *Researching Mathematics Learning*. Details of this and other Open University courses can be obtained from the Student Registration and Enquiry Service, The Open University, PO Box 197, Milton Keynes MK7 6BJ, United Kingdom. Tel: +44 (0)870 333 4340. e-mail general-enquiries@open.ac.uk

Alternatively, you may visit the Open University website at http://www.open.ac.uk where you can learn more about the wide range of courses and packs offered at all levels by The Open University.

To purchase a selection of Open University course materials visit the webshop at www.ouw.co.uk, or contact Open University Worldwide, Michael Young Building, Walton Hall, Milton Keynes MK7 6AA, United Kingdom for a brochure. Tel. +44 (0)1908 858785; Fax +44 (0)1908 858787; e-mail ouwenq@open.ac.uk

Published by Tarquin Publications; written and developed by The Open University

Tarquin Publications
99 Hatfield Road
St Albans
AL1 4JL
www.tarquinbooks.com

The Open University
Walton Hall
Milton keynes
MK7 6AA
www.open.ac.uk

This edition published 2006

British Library Cataloguing-in-Publication Data available on request.

Paper ISBN 1899618651
Cased ISBN 1899618678

Typeset by DPSL.
Printed and bound in the United Kingdom by Lightning Source

ACKNOWLEDGEMENTS

The authors would like to thank the following for their part in making the publication of this book possible:

Eric Love, for thoughtful academic editing;

Kate Richenburg, for experienced editing;

Sally Crighton, Peter Johnston-Wilder and Ron McCartney for careful critical reading.

Laurinda Brown for feedback

Andrew Whitehead, for the artwork

The many teachers and colleagues who have participated in the development and refinement of these articulations.

CONTENTS

PREFACE

This book was written as part of a course in Researching Mathematics Learning (ME825) at The Open University. It summarises an approach developed in the Centre for Mathematics Education over 25 years, which involves labeling past experiences in order to bring to mind pedagogical choices, which in turn inform the practice of teaching mathematics.

In order to promote the development of teaching practices through informing choices, the focus of the book is on the design and use of tasks. The heart of teaching is interaction with learners; activity by learners creates opportunities for pedagogically effective interaction; learner activity is initiated by tasks; tasks are chosen so as to enable learners to encounter significant mathematical ideas and themes, and to enable them to use their own powers in making sense of mathematics and mathematical sense of their world. Hence the title of the book.

John Mason is Professor of Mathematics Education at The Open University, where he was, for some 15 years, Director of the Centre for Mathematics Education. He has a lifelong interest in thinking mathematically himself, in working with others who want to think mathematically, and in supporting people who want to work with others to think mathematically.

Sue Johnston-Wilder is a Senior Lecturer in Mathematics Education at The Open University. She has worked with teachers and student teachers for many years developing materials to promote interest in mathematics teaching and learning.

1
INTRODUCTION

Mathematics seems, on the surface, to be a cut and dried subject. However, it is very difficult to decide just what is meant by being able 'to do mathematics'. Does it mean mastery of techniques? Does it mean appreciating how ideas and techniques link together? Does it mean being able to reconstruct facts and techniques from basic principles? Does it mean knowing when it is appropriate to use one technique rather than another? Does it mean being good at solving novel problems? If being able to do mathematics involves all of these, how can *learning* mathematics be described?

In this book, learning is taken to involve some sort of transformation in the way that learners perceive or think. This transformation could involve extending learners' sensitivities and awarenesses, as well as increasing the choice of actions to which learners have access. In the process of transformation, learners will develop their powers to think mathematically, their competence and fluency in using specific techniques and language, and their appreciation of how ideas fit together.

> The word 'learners', rather than 'children', 'pupils' or 'students', is used throughout this book to include learners of all ages.

1.1 ACCEPTING AND ASSERTING

How then do teachers help learners to achieve such a transformation?

This book starts with the premise that learners must be active. It describes ways in which teachers can encourage learners to use their initiative rather than just be passively present in the classroom.

The crucial factor in this context is the distinction between *accepting* and *asserting*. When learners sit back and think to themselves 'I could have done that', or when they wait to be told exactly what to do, they are in an *accepting* mode. When they ask probing questions and use their initiative, they are in an *asserting* mode. In an asserting mode, learners take risks and are ready to learn from their mistakes; they also make and test conjectures and try to reconstruct ideas and techniques for themselves.

For a long time it has been advocated that learners need to be active if they are to learn effectively. Herbert Spencer, an educational reformer of the early twentieth century, claimed that:

> Rule teaching is now condemned as imparting a merely empirical knowledge—as producing an appearance of understanding without the reality. To give the net product of inquiry, without that inquiry that leads to it, is found to be both enervating and inefficient. General truths to be of due and permanent use, must be earned.
>
> Spencer, 1929, p. 57.

The mathematics educator Richard Skemp (1976, 1979) drew a distinction between *instrumental understanding* (based on memorisation of the steps of mechanical procedures) and *relational understanding* (the thoughtful and connected learning of principles); these two kinds of understanding mirror the ideas of accepting and asserting. Accepting behaviour is, at best, likely to lead to instrumental understanding, whereas asserting behaviour is likely to promote relational understanding. However, it is not easy to distinguish between 'accepting' and 'asserting' merely by observation. What may appear outwardly to be accepting behaviour may mask inner asserting, and what may appear to be asserting behaviour may be mimicry of the form but without any actual inner transformation.

Becoming aware of the distinction between *accepting* and *asserting* behaviour can prompt teachers to look for opportunities to encourage learners to take the initiative. It is one of a number of such distinctions that have proved fruitful in the design and use of mathematical tasks. In this book, various distinctions that aim to generate useful activity and enable fruitful interactions between teacher and learner are considered; for example, it is important to distinguish between the behaviour, emotion and awareness of the learner.

1.2 BEHAVIOUR, EMOTION, AWARENESS

John Dewey, the American philosopher–educator of the nineteenth and early twentieth century who was latterly influenced by Herbert Spencer, suggested that an important preliminary to lessons is *psychologising the subject matter*; that is, structuring tasks so that they give learners access to the themes, processes and specific features of a topic. The teacher needs both to be involved in the pyschologising process, and to have assistance from outside, especially from curriculum designers. For instance, in one approach to psychologising the subject matter of school mathematics topics, the Dutch mathematician and mathematics educator Hans Freudenthal (1983) used phenomena that every learner would have experienced (such as shadows, turning the body, the use of adult-sized furniture by small children and so on), which consequently a teacher could invoke for immediate discussion.

To psychologise the subject matter, one needs some idea of the structure of the human psyche. There is an ancient and fundamental division of the psyche into various components which overlap, intertwine and mutually sustain each other. The words *behaviour, emotion* and *awareness* are used to identify three important components of the psyche that represent, respectively, the mind acting on the senses, on the emotions and on itself. In a variety of forms these components pervade this book, and so it is useful to explore them more fully.

- *Behaviour* is what one does observably in the material world. Later we shall generalise 'behaviour' to refer to the manipulation of objects, images, symbols and language.
- *Emotion* is the emotional response to a situation, and also the source of the energy that drives learners and teachers.
- *Awareness* is associated with intellect; that is, with what one is explicitly and implicitly aware of or sensitised to.

In another, related, approach, Bruner (1966) drew attention to three modes of representation: *enactive* (doing things), *iconic* (images and sense-of) and *symbolic*. Thus, when considering a paper cup rolling around on the floor, you could actually roll the cup and observe the path that it follows (enactive mode); you could imagine the cup rolling, and use a diagram to locate and relate relevant quantities (iconic mode); you could use symbols to represent different lengths and then formulate equations that relate the lengths to one another (symbolic mode).

Another illustration of Bruner's modes concerns areas of rectangles. Learners can 'count squares' to find the areas of rectangular figures and of figures made up of rectangles (enactive mode). They can also become aware that counting squares can be done more efficiently by multiplying (iconic mode). Multiplication can then be thought of in terms of calculating rectangular areas, even when the specific values are unknown and are represented by letters (symbolic mode). Images of rectangles can help when expanding brackets, when factorising and when linking to the distributive law as an expression of mental strategies for multiplication ($7 \times 34 = 7 \times 30 + 7 \times 4$). Eventually the images can fade into the background as learners become competent at multiplying numbers and symbols.

Bruner's three modes can be related to the three aspects of the psyche:

- *Enactive:* People do things, and the doing provides knowledge about the world (behaviour).
- *Iconic:* People imagine, and through their imagination they direct and control their emotions.
- *Symbolic:* People use symbols to signify things not present and employ abstractions that cannot be presented in the material world (awareness).

The *enactive* mode of representation particularly applies to the world of material objects, but it also applies to anything that can be manipulated with confidence. The *iconic* mode relates to the world of images and 'having a sense-of'. The *symbolic* mode corresponds to the world of symbols that are displaced from what they represent and isolated from their original context. These three modes are overlapping and intertwining, rather than distinct and separate. Some people make the mistake of thinking that they form a hierarchy with 'symbolic' as the goal. However, each mode enriches and is enriched by the others. Thus, symbols can have iconic associations as well as triggering memories of physical actions. What is initially 'something people do' gradually becomes an object, and therefore a thing that people can act upon.

One example of this transformation from 'doing' to object is provided by number. The learning of number begins with the action of counting (three pencils, two cubes); gradually the numbers become objects in their own right (three, two) and are abstracted from context. Through repeated application, these abstract objects become familiar and are used as confidently as if they were material objects. The numbers are then acted upon (the operations of arithmetic), and these operations become things (addition, subtraction, and so on) in an ever-expanding spiral of abstraction and symbolisation.

The term *reification* is used to describe this shift in which a process is treated as an object. As this shift is so central to developing mathematical competence and confidence, it has been much studied and will be considered more fully in Chapter 6.

1.3   MATHEMATICAL TOPICS, TASKS AND ACTIVITIES

The basic aim of a mathematics lesson is for learners to learn something about a particular topic. To do this, they engage in tasks. By 'tasks' we mean what learners are asked to do: the calculations to be performed, the mental images and diagrams to be discussed, or the symbols to be manipulated. In this book we shall ignore other 'tasks', such as 'listen to me' or 'give out the books', as well as tasks that are devised solely for assessment rather than for promoting learning.

Although the words 'task' and 'activity' are often treated as synonyms, we follow Christiansen and Walther (1986) in drawing a distinction between them. The purpose of a *task* is to initiate *activity* by learners. In such activity, learners construct and act upon objects, whether physical, mental or symbolic, that pertain to a mathematical topic. This activity is intended to draw learners' attention to important features, so that they may learn to distinguish between relevant aspects, or recognise properties, or appreciate relationships between properties. Tasks are the subject of Chapter 3, and learner activity of Chapter 4.

Learners are not just passive recipients of skills and facts. They are innate learners and bring to lessons their natural powers of sense-making. The tasks they undertake need to offer an appropriate challenge. But merely doing the tasks is not enough, because while carrying out a task, learners may not stress what the teacher sees as the important experiences, and so they may not integrate relevant experiences into their functioning. This means that, although some learning will certainly take place, it is difficult to guarantee that what is learned is what the teacher intended to be learned. For instance, Denvir and Brown (1986a, 1986b) found to their surprise that, in the short term, learners appeared to learn, or at least to get better at, techniques and topics that were not even part of the sequence of lessons.

Also, when learners have a sequence of experiences, it does not follow that they actually perceive it as a *sequence* rather than as a disjointed collection. Yet in order to learn what was intended, they need to be aware of the connections between the experiences. Hence, for learners to get a sense of what is meant by 'an angle', they need to put together a number of related awarenesses. These include: appreciating that an angle is an amount of turning, that the position of the vertex and the length of the arms are irrelevant, and that the angle has a sign depending on the positions of the 'initial' and 'terminal' arms relative to anti-clockwise motion around the vertex. The learners need multiple experiences, together with reflection on those experiences, to draw out what features are invariant and what can change if two angles are described as 'the same'.

Overall, learning is best seen as a maturation process, which necessarily takes time. Although some learning will take place naturally, it can be accelerated through explicit reflection on experiences. Put another way:

teaching takes place *in* time; learning takes place *over* time.

Perhaps the central issue in teaching mathematics is how to support learners in making mathematical sense of their experiences. At the heart of these experiences is learners' *activity,* which occurs as the result of undertaking *tasks* that are informed by mathematical *topics.* These aspects are explored in subsequent chapters of this book.

1.4   TEACHING TO SUPPORT LEARNING

The very earliest written records from Babylon, Egypt and China include mathematical texts with problems and worked examples. This suggests that people have always appreciated the need for learners to be active: to follow worked examples and to do exercises for themselves. Plato, for example, extolled the Egyptian practice of using physical objects to introduce number, while criticising the Greek practice of not using them (Hamilton and Cairns, 1961, pp. 353–84).

Over time, a variety of teaching practices have been tried and, in isolation, found wanting. Some of these are:

- presenting a rule and getting learners to memorise that rule before applying it in particular cases;
- starting with worked examples and then deriving 'rules' from these, or getting learners to derive the rules from the examples for themselves;
- starting from the general case and moving to the particular;
- starting from the particular (either simple or complex) to derive the general.

No single strategy has proved universally successful. This book advocates a 'mixed economy', in which learners are given a variety of types of task to develop mathematical thinking. It sets out in detail some of the issues and choices surrounding the design and implementation of tasks and draws upon many sources and theories. We start from four fundamental notions:

- learners need to be mentally, emotionally and sometimes, even, physically active;
- teachers need to be aware of relevant features of the topic to be learned, of the mathematical structure of the topic, and of ways of 'psychologising the subject matter' so as to make it accessible to students;
- tasks become vehicles for learning because of the quality and nature of the interactions and activities arising from them, and the ethos in which the tasks are undertaken;
- learning takes place within a setting, that is, an environment of social forces and classroom practices (called a *milieu* by Brousseau (1997), see Figure 1.1).

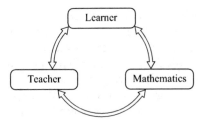

Figure 1.1 Teacher–learner–mathematics within a milieu.

The approach in this book is based on *activity theory,* which was developed particularly by Vygotsky (1978) and Leont'ev (1981a), and extended by Wertsch (1981), Bruner (1986) and others. According to activity theorists, the mind develops, and can only be understood, within a context of inter-action with other people and with the environment. One implication of this approach is that tasks need to be designed so that learners have:

- relevant experiences from which to extract, abstract and generalise principles, methods, perspectives and ways of working with mathematics;
- stimuli appropriate to the concepts to be worked on;
- a supportive and compatible social environment in which to work.

As Vygotsky taught, learners also need access to adults and experienced peers who exhibit the kind of thinking that learners are expected to acquire.

Overall, what matters most are the 'ways of working', not the mathematical concepts or the tasks. Consequently, your own experiences of learning and teaching mathematics are important, so consider the following three questions before you read the next section:

- What did you personally find helpful to do in order to learn mathematics effectively?
- Based on your experience, what do you think that the learners you teach need to do in order to learn mathematics effectively?
- Based on your experience, what do you think that the learners you teach believe that they need to do in order to learn mathematics effectively?

1.5 EXPERIENCE

When people become teachers they develop strong and deeply held images of what learners need to do, but these are not always consonant with what the learners themselves think or need, nor even consonant with the teacher's own past experience.

Learners draw on their own recent experience of mathematics, which may not have been positive or effective. Many learners in school act as if they think that managing to get the correct answers to most of the tasks that are set by their teachers fulfils their side of the bargain—their part of the implicit 'contract'—and so assume they will learn what is required. If only it were that simple! It is so easy for learners to listen passively, accept what is said, follow instructions, and believe that they are learning. This may be in spite of repeatedly finding themselves at a loss when they are asked to assert something, whether by 'doing some exercises', asking questions or making conjectures.

Have you noticed that some learners do not do the task you set or, at least, the task you thought you set? Have you asked learners at the end of a lesson to write down what they thought the lesson was about and been surprised by the responses?

Jaworski (1994) questioned learners about their mathematics lessons. After a lesson in which learners cut out a quadrilateral shape and then tried to tessellate a plane with it, some of them said that they thought the lesson was about cutting out, some that it was about using scissors, some that it was about quadrilaterals. Very few recognised that it was about tessellation, and even fewer realised that there was an underlying result: that every quadrilateral tessellates! So what learners get from their activity is highly idiosyncratic.

Stein (1987) distinguished between the task as set by the teacher and the task as interpreted by the learners. This idea can be extended, with more distinctions added:

- the task as imagined by the task author;
- the task as intended by the teacher;
- the task as specified by the teacher–author instructions;
- the task as construed by the learners;
- the task as carried out by the learners.

These complexities are discussed in more detail in Chapter 3.

Teachers need to interpret tasks designed by other people and need to form views on what the intentions of the author were and what preparation the learners need if they are to benefit from undertaking the task. Teachers also need to be aware of how tasks might be interpreted by learners and know how to support the learners' activity in such a way that the experience can result in appropriate learning.

1.6 DIMENSIONS-OF-POSSIBLE-VARIATION

We examine the design and development of tasks by looking at the ways in which tasks can be varied without losing their purpose. As an example, consider the following tasks which involve specific numbers, and think about how these numbers might be changed without changing the 'mathematical operations'.

Task 1

Write down a pair of numbers such that one number is over 50 and the other is between 30 and 50. Subtract the smaller number from the larger one. Form a new pair of numbers consisting of the smaller number and the difference. Repeat until you get a difference of zero. The number just before you get zero should be the greatest common divisor of your starting numbers.

Task 2

Write down a decimal number lying between 2 and 3 but not using the digit 5, which does use the digit 8 and which is as close to $2\frac{1}{2}$ as possible.

Task 3

Simplify the following fractions: $\frac{21}{49}, \frac{12}{28}, \frac{18}{42}$.

Task 4

A horse-drawn carriage used in parades has back and front wheels of different sizes, as shown in Figure 1.2. The large back wheels are 165 cm in diameter, and the smaller front wheels are 121 cm in diameter. If the positions on the wheels where they touch the ground at a particular moment are marked so as to be easily visible, how far does the carriage have to travel before these marks are all touching the ground at the same time again?

Figure 1.2 A horse-drawn carriage with wheels of different sizes.

In Task 1, any pair of numbers within a given range will do (but could one or both of them be negative or fractional?). In Task 2, the number $2\frac{1}{2}$ could be changed suitably to produce a similar task, as could the other digits that are specified. In Task 3, all of the fractions are equivalent, so there are some implied restrictions in choosing other fractions. Task 4 can have any positive numbers but is mathematically interesting only when the two numbers share a common divisor.

The fact that numbers can be changed without affecting the method of solution or altering the concepts involved points to a *dimension-of-possible-variation* in the tasks. (This notion is based on an idea of Marton (see Marton and Booth, 1997).) For each number that could be altered, there is an associated *range-of-permissible-change* which takes account of any necessary restrictions or constraints within the task. For instance, in Task 1 the numbers need to be positive integers, in Task 4 the wheels of the carriage need to have positive diameters, and so on.

Altering the numbers in a task is the most obvious of several dimensions-of-possible-variation that transform a task from a single exercise into a class of problems or a 'problem type'. Learners make progress when they become aware that what matters about a task is the method and the thinking involved, rather than the specific numbers. When they begin to think about a problem type, they are starting to think mathematically *about* tasks as well as *within* tasks.

There are other dimensions-of-possible-variation to consider. For example, the context of Task 4 could be changed. It could become a task about two (or more) people walking along the same track at the same speed but taking different-sized paces. Further variations could involve people walking at different speeds, or walkers going in opposite directions around the track. So the context can be changed and implicit numbers such as the number of people can also be changed, while not affecting the method of solution or the concepts involved.

Different ways in which tasks can be formulated and presented are considered in Chapter 3, where the notion of dimensions-of-possible-variation is applied both to the tasks themselves and to features within the tasks.

Meanwhile, important questions for teachers—and learners—to keep in mind when engaging in tasks are:

- What aspects of the task are generic and, hence, fixed?
- What aspects are particular and can be changed?

Put more succinctly, looking for *invariance* in the midst of *change* can enrich learners' experience of tasks, as well as inform and guide mathematical exploration.

van Hiele (1986), along with Maturana and Varela (1988), claimed that the ability to make distinctions—to distinguish one aspect from others—is vital to human and animal functioning (see also Hauser, 2001). Humans naturally seek and detect similarities and differences between juxtaposed experiences. Therefore it is important that tasks involve learners in making distinctions. This kind of involvement is most readily brought about by systematically varying one dimension-of-possible-variation while other aspects stay the same.

Marton's view (Marton and Booth, 1997) was that learners need to experience systematic variation in each relevant dimension in rapid succession, so that these experiences of particular instances are juxtaposed and compared through reflection. Only then are these experiences likely to be appreciated and generalised. If too many things are varied at once, then it is hard to make individual distinctions; on the other hand, if particular instances are too far apart in time, learners may not link them together and so may not see them as examples of variation in some dimension. Hence the power of the question 'What is the same and what is different ...?' for prompting mathematical activity.

1.7   THEORY AND PRACTICE

He who loves practice without theory is like the sailor who boards ship without a rudder and compass and never knows where he may cast. Practice always rests on good theory.

Attributed to Leonardo da Vinci.

We would like our theories to be as fact-based as our facts are theory-based.

Goodman, 1978.

In discussions of teaching and learning, it can be tempting to stress experience and practice but to deny any role for theory. However, as the quotations above suggest, behind every effective practice there is some sort of theory. What constitutes a 'good theory' is, of course, open to question, but we shall take it to mean a collection of guiding principles that inform practice. Theories do not need to be grand philosophical systems: they can be simple assumptions which, although they inform practice, may never have been made explicit. Theories lie behind sentiments such as 'learners need plenty of practical experience' or 'teachers need to state things clearly'. People perceive and make distinctions by means of 'theories', however hidden those theories may be. Furthermore, for teachers to develop their practice, they need to bring implicit theories to the surface so that those theories can be examined and, perhaps, modified.

A useful concept in relation to theories is that of *frameworks*. A framework is a coherent set of ideas that together emphasise important aspects of a phenomenon, and that constitute an overt theory that can be put to practical use. For example, the traditional suggestion to teachers, sometimes called the 'Three T's'—'Tell them what you are going to tell them, tell them, and then tell them you've told them'—constitutes a very simple framework for devising a lesson plan. No single framework can capture all aspects of teaching and learning, but it can act as a reminder of several strands and can also be very helpful when discussing teaching and when planning lessons. Different frameworks are rather like different photographs of an object, each giving a distinctive view. In any given framework, some features will be revealed, while others remain hidden.

You will find that you sometimes agree and sometimes disagree with the theories we present and the underlying assumptions. Disagreement is an easier sentiment to harness, because you can check what is said against your immediate experience. It is much harder to check what you already find compatible and harmonious. Yet it is only by questioning the familiar that professional development takes place.

Some of the assumptions underlying the theories of learning that we put forward concern the kinds of transformations that learning requires. These transformations include:

- becoming aware of learners' innate or natural powers to think in certain ways, or apply those powers in new ways;
- becoming more sensitised to making distinctions between things that were previously blurred, as in learning new concepts;
- developing and honing new skills and techniques;
- becoming aware of how ideas and techniques are connected, perhaps by recognising the presence of core mathematical themes;
- having new facts at one's fingertips.

In order to experience such transformations, learners need to become active—to move from accepting to asserting, and from doing things to making sense of them.

Looking ahead, in Chapter 2 we shall examine the structure of mathematical topics in order to feed task design. Readers eager to work on tasks may choose to skip ahead to Chapter 3, but will necessarily find themselves asking what the tasks are intended to achieve, and this will draw them back to the structure of mathematical topics. Readers keen to work on learner activity can turn to Chapter 4, but they will also soon find themselves drawn to consider different ways in which tasks can be structured; they will then want to look at the aim and purpose of those tasks—aspects which again relate to the topic structure. In Chapter 5 we shall examine the verbal interactions that help to initiate and sustain activity, and in Chapter 6 we stand back and look at pupil progression.

SUMMARY

The heart of teaching lies in interaction with the learner, with the aim that fruitful learning will take place. Interactions need purpose and context, and these are provided by learner activity. Activity is initiated by tasks that are constructed in such a way that the learner can encounter important ideas in curriculum topics.

The structure of this book (see Figure 1.3) reverses this path and goes from topics, through tasks, to sustaining activity through interaction between teacher and learner. The outcome is progression in thinking mathematically and in developing competence in specific topics.

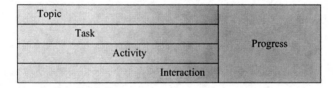

Figure 1.3   Structure of the book.

Three useful ideas, which will help to structure the remaining chapters, have been introduced. These are *frameworks, dimensions-of-possible-variation* and the associated *range-of-permissible-change*.

Frameworks are helpful as ways of informing the design and presentation of tasks and promoting the effective use of those tasks. Frameworks embody distinctions, some of which have already been introduced:

- Accepting–Asserting;
- Behaviour–Emotion–Awareness;
- Enactive–Iconic–Symbolic.

The idea of dimensions-of-possible-variation has been described in connection with tasks. Tasks can be varied along any of several dimensions to become representatives of a general class of tasks. Most importantly, learners need to appreciate dimensions-of-possible-variation and the corresponding range-of-permissible-change in order to comprehend and use a definition, result or technique.

2

MATHEMATICAL TOPICS

This chapter introduces a framework for thinking about mathematical topics. The framework will provide a background against which to consider the nature and use of mathematical tasks when you read Chapter 3.

The underlying notion of this chapter is the three-component structure of the psyche introduced in Chapter 1: behaviour, emotion, awareness. These components are used to examine how new topics arise, and are then incorporated into the structure of any mathematical topic, thereby providing a teacher with a way of preparing to teach a topic.

2.1   WHERE DO MATHEMATICAL TOPICS COME FROM?

A mathematical topic arises when people become aware that they are able to perform an action—in other words, to do something. In the midst of performing the action, they become aware of what it is that they are doing, and so they step out of the 'doing' itself. The 'doing' then becomes a thing that can be formally studied.

A typical 'doing' occurs when someone manages to solve a class of problems using a single technique. The technique is then extracted from its context and refined for teaching to learners of a particular age or experience. Often the originating problem or problems are obscured or forgotten. In this way, methods for performing arithmetic operations were abstracted from many contexts and became context-free computations to be taught and mastered. The arithmetic of fractions, decimals and percentages emerged in this way. Topics such as the solution of equations, factorisation of quadratics, areas of shapes, Pythagoras' theorem and, indeed, every other topic in school mathematics arose similarly, that is as a means of solving a class of problems. Gattegno (1987) proposed that every topic arises in the same way; that is, from people becoming aware of actions that they can perform, so the actions themselves become the object of further study and of further actions. For instance, mechanics begins from the awareness of pushing and pulling, algebra from the awareness of the manipulation of relationships, and geometry from the awareness of the dynamics of mental imagery.

Seeing topics arise from classes of problems suggests that the motivation for a topic can often be found in versions of the original problems that the topic resolves, and in the range of problems to which that topic can be applied in different contexts.

2.2   STRUCTURE OF A TOPIC

Perhaps the most familiar way of thinking about a topic is as a collection of techniques to be mastered, rather like a collection of tools. Learners show that they have mastered the topic by displaying competent behaviour with the relevant techniques in appropriate circumstances. A topic can therefore be looked at in terms of the behaviour associated with it.

However, in order to gain facility and competence with techniques, it is necessary to become familiar with technical terms and their definitions, as well as with facts and theorems; it is also essential to recognise the classes of problems that those techniques can resolve.

CONCEPTS AND CONCEPT-IMAGES

We start by considering the concepts that underlie a topic. But what is a 'concept'? It sounds like a 'thing', but it cannot be pointed to, possessed or packaged in the way that a material object can. Rather, a concept is a label for the flow of images, thoughts, sensitivities, connections, possible actions and so on associated with an idea.

Concepts are not isolated entities floating about in our minds. Rather they are familiar 'lines of thought'. Piaget (1971) used the notion of *schema* to refer to the web of interconnections associated with an idea. Skemp (1976, 1979) developed this within mathematics education.

For an expert, the mere mention of a technical term in mathematics, such as 'angle' or 'mean', gives access to a variety of associations, techniques, ways of speaking, images, symbols and meaningful contexts. Tall and Vinner (1981) captured this experience in the term *concept-image*, which encompasses the whole mental structure associated with a concept. A concept-image is:

> the total cognitive structure that is associated with the concept, which includes all the mental pictures and associated properties and

processes. It is built up over the years through experiences of all kinds, changing as the individual meets new stimuli and matures.

Tall and Vinner, 1981, p. 152.

An essential feature of the concept-image is that the various aspects are interconnected. Thus, for some people, the term 'repeated subtraction' may trigger the idea of division. But this is unlikely to happen with a learner who is not yet confident with division, because, in general, people's images and attention tend to be centred around behaviours and concepts about which they are confident. Similarly, the term 'division' may trigger the concept of repeated subtraction, and also fractions, decimals, long division, sharing and a host of other notions. To pick up on what learners are thinking and doing, and to prepare tasks that will provide learners with access to the connections within a topic, teachers may find it useful to refresh their own versions of these connections.

As a further example, consider the words 'Euclidean algorithm'. These words may not have any meaning at all for you, except for a vague association of 'Euclidean' with geometry. If the term 'greatest common divisor' is included, you may now be triggered into recalling Task 1 in Chapter 1, and this may, in turn, trigger some recollections of doing things with pairs of numbers. There might also be some recollection of repeated subtraction as the basis of the method by which you might reconstruct a technique for finding the greatest common divisor of two numbers. This technique is, in fact, called the 'Euclidean algorithm'. In a different context you might find more, or fewer, links coming to mind. If you had been using the algorithm recently, you would be more likely to have richer connections than if you had not used it for a long time. Everything that comes to mind is part of your concept-image of the 'Euclidean algorithm'.

It is not possible to gain direct access to someone else's concept-image. Furthermore, people in different circumstances and with different experiences will bring to mind different associations and connections. Nevertheless, the notion of a concept-image is useful for a teacher or author when designing tasks. Teachers, by reminding themselves of the various connections, associations, images, techniques, language patterns, examples, results and proofs associated with a topic can identify the features of the topic that they want learners to encounter and to internalise. The rest of this section systematises these ideas and extends them to provide a framework for the structure of any mathematical topic.

A FRAMEWORK FOR TOPICS

Consider, for instance, the notion of *randomness*. What immediate associations come to mind? Do you recall some specific examples, or if you were given examples, could you say in what way and to what extent you consider the word 'random' to be appropriate? What other words would you use in connection with 'random'? What techniques do you associate with the word? The ideas, images and connections that present themselves to you form your current concept-image. Of course, different situations may trigger different aspects, which will all contribute to your overall concept-image.

The questions just posed in relation to randomness are generic: they apply to any topic that a teacher may be preparing to teach. Indeed, the questions are representative of a much larger set that can usefully be split into three categories:

- Questions concerned mostly with *behaviour*; these deal with what learners can say and do (that is, with language and techniques).
  Generic examples of such questions are:
  What special words and phrases are associated with the topic? How are those or similar words used in ordinary language? What techniques or procedures form part of the topic? What technical vocabulary is involved, and what are the phrases and sentences that learners need to begin to use in order to speak cogently about the topic and to use the techniques? What conversations inside the head support the use of the techniques?
- Questions concerned mostly with *emotion*; these deal with what sources and occurrences of topic ideas motivate learners, and what surprises and disturbances they encounter.
  Generic examples of such questions are:
  What problems or problematic situations arouse interest and develop into methods that become the topic as you now know it? In what sorts of context might the ideas and techniques appear? What sorts of problem does the topic help to answer? Where does the topic arise, or in what sorts of situation is it likely to arise? What surprises, dissonances or disturbances might be encountered?
- Questions concerned mostly with *awareness*; these deal with what learners attend to and experience inwardly, and what they might recall in order to access appropriate behaviour.
  Generic examples of such questions are:
  What images, intuitions, sense-of, connections and associations come to mind? What sorts of representation are relevant and informative (whether physical objects, diagrams, charts, tables or symbols)? What are useful generic examples? What sorts of complication and difficulty tend to arise? What themes might be encountered? What links enrich the topic or concept? What are the underlying structures or principles from which the topic is derived? What do learners need to attend to specifically?

Answers to questions such as these are partly personal, and the usefulness of an individual question will vary from concept to concept. Together, all of your answers constitute your version of the topic—your concept-image. We suggest you pause now, take a different topic, such as fractions or angles, and think how these questions would apply.

A succinct way to display the three categories of questions outlined above is as three interwoven threads which make up the framework that we call the 'structure of a topic' (see Figure 2.1). This can assist teachers in selecting, augmenting and designing tasks that contribute to learners' experience of a particular topic. The horizontal thread relates to emotional factors; the behavioural thread runs from top-left to bottom-right; the awareness thread runs from bottom-left to top-right.

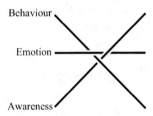

Figure 2.1   Structure of a topic.

Gattegno (1987) asserted that *only awareness is educable*. This statement makes a strong claim and is intended to be challenging. Corresponding statements about behaviour and emotion, which may help to put Gattegno's observation into context, are: *only behaviour is trainable*, and *only emotion is harnessable*. Behaviour is what is trained, but if behaviour is to be flexible and responsive to changing situations, then it must be guided and directed by awareness, that is, by sensitivity to subtle differences. Thus awareness is the aspect of the mind that can be educated. Emotions such as surprise, disturbance and cognitive conflict, together with the pleasure of using one's intellectual powers, provide the motivation for this education.

The Behaviour–Emotion–Awareness framework can be used to inform the kinds of questions that teachers ask themselves when they are preparing a topic to teach. The framework is of limited value as a mechanical checklist, but it can be informative as a structure to refer to when uncertain whether a topic has been fully covered. As the basis for an on-going collection of notes on a topic, it can be invaluable.

With regard to making notes, every time a topic is taught, the teacher can look out for unusual responses from learners and jot these down. They might arise in relation to:

- words and phrases that the teacher had not previously appreciated might confuse learners (for instance, words with everyday non-technical meanings, such as diamond, sum and differentiation);
- what is said or done that shows a learner to be in the midst of sorting things out but not yet having a clear understanding;
- phrases or mnemonics that the teacher or the learners use to guide the application of a technique;
- unusual contexts in which the techniques are applicable or the ideas are useful;
- new problem situations that might give rise to a need to learn about the topic.

When a teacher prepares to teach the topic again, a collection of such notes is likely to be more helpful in alerting the teacher to important issues than a previous lesson plan specific to a particular class.

2.3   BEYOND PARTICULAR TOPICS

Mathematics is traditionally divided into domains such as arithmetic, algebra, geometry, and data handling. Topics are often allocated to one of these domains, as follows:

- *Arithmetic:* Adding and subtracting; multiplying and dividing whole numbers; fractions; negative numbers; decimals.
- *Algebra:* Expanding brackets and factorising; setting up and solving linear equations in one or more variables; solving quadratics; completing the square in order to obtain the formula.
- *Geometry:* Angle properties of triangles and quadrilaterals; naming shapes; circle theorems; triangle theorems.
- *Data handling:* Mean; median; mode; standard deviation; distributions.

There are two problems with this approach. Firstly, some topics do not fit easily into one of the categories. For example, coordinate geometry is both algebra and geometry; similarly, measures are used in both arithmetic and geometry.

The second problem is that, by separating topics in this way, mathematics can degenerate into a large collection of disparate techniques and vocabulary, with connections between the topics being ignored. The result is that closely related concepts such as decimals and fractions are seen as distinct entities, and intimate connections between topics, such as that between straight lines and ratios, are completely overlooked. Many textbooks accentuate this disconnection by the way they organise different topics on successive pages 'so learners won't get bored'. However, Askew *et al.* (1997) looked at teachers of classes which do well in standard assessment tasks. They found that the most significant feature of the teachers was the richness of the teachers' awareness of how mathematical topics fit together.

No matter how mathematics is divided into domains and topics, there will be features that are omitted. Typically, aspects that cut across all categories, such as problem-solving and reasoning, or communicating using mathematics, do not belong to particular topics although they are increasingly seen as important aspects to be developed at school.

Rather than trying to fit topics into categories, it can be helpful to think in terms of some pervasive mathematical themes that serve to unify topics which might otherwise appear to be disparate.

MATHEMATICAL THEMES

Four of the most important mathematical themes are: 'freedom and constraint', 'invariance in the midst of change', 'extending and restricting meaning' and 'doing and undoing'.

Freedom and Constraint

Consider Task 2 from Section 1.6:

> Write down a decimal number lying between 2 and 3 but not using the digit 5, which does use the digit 8 and which is as close to 2½ as possible.

Here you are free to choose a number, but because there are conditions attached, your freedom is subsequently constrained. Similarly, when you draw a triangle, the first angle can have any value between 0° and 180°, and the two other angles will be even more constrained.

It is useful to appreciate that an equation is a constraint placed on an unknown which, without that constraint, has the freedom to take any value at all. Moreover, every mathematical problem can be seen as a task to construct an object that has its freedom restricted by explicit and implicit constraints. Sometimes the freedom is so fully restricted that no construction is even possible; take, for example, the problem of where the lines $y = 3x + 2$ and $2y - 6x = 10$ intersect. Most tasks can be thought of as starting from objects with great freedom and then gradually restricting that freedom through the imposition of a sequence of constraints.

Invariance in the midst of change

Every mathematical concept or definition involves specifying something that can change while something else remains invariant. For instance, the equivalent fractions $\frac{1}{2}, \frac{2}{4}, \frac{3}{6}, \frac{4}{8}, \ldots$ all represent the same number, even though the numerators and denominators are different: that number, 'one-half', is an invariant. Similarly, the size of an angle is invariant regardless of the length of the containing arms or the position of the vertex. Most mathematical results or facts are statements about something remaining invariant while something else changes. Thus, the sum of the angles of a triangle is always 180°, or two right-angles, or a half-turn (the units are unimportant), no matter how the triangle's shape changes.

When a mathematical task consists of a set of similar questions, there is usually an invariant present: it might be that the questions are all 'of the same kind' or that they all 'use the same method'. Other tasks might require learners to discover and express something that is the same about a class of objects; for example, that any two odd numbers added together will produce an even number, or that all quadrilaterals tessellate the plane. In these examples, the words 'any' and 'all' indicate that there is an invariant.

Extending and Restricting Meaning

The term *number* is used for the counting numbers 1, 2, 3, ... , and then extends its meaning: first, to include zero, then to include fractions and decimals, then to include negatives, and later to include other kinds of 'number'. At each stage what is encompassed by 'number' includes new objects, although they all satisfy the properties of arithmetic operations. Similarly, the term 'square' becomes a special case of a rectangle as well as of a kite, trapezium, rhombus, parallelogram and quadrilateral. Another example is the extension of the notation for powers of a number, $3^2, 3^3, 3^4, \ldots$, to include 3^0, 3^{-1} and so on.

Sometimes meaning is restricted. Isosceles triangles are a restricted class of triangles that have a pair of sides equal and other specific properties which can be inferred. Those numbers that leave a remainder of 1 on dividing by 3, {1, 4, 7, 10, 13, ...}, are a restricted set of whole numbers and also have particular properties; for example, when any two such numbers are multiplied, they produce another number in the set. There are numbers in the set that take the role of prime numbers, but are not always prime numbers in the usual sense: in this case, 10 takes the role of a prime number because no other number in the set divides into it, whereas 28 does not because it is 4 × 7. So restricting the object can lead to using old terms, such as 'prime', with new meanings.

Doing and Undoing

Whenever you can do something (for instance, add or multiply two numbers to get an answer, find the area of a figure made up of rectangles, or expand brackets), there is a reverse question waiting to be asked: 'If this is the answer, what could the question be?' Thus, $3 + 4 = 7$ turns into $7 = ? + ?$, and $3 \times 4 = 12$ turns into $12 = ? \times ?$ More generally, the reverse of multiplying two numbers together is to split a given number into factors; the reverse of expanding brackets is to factorise an algebraic expression. Such reversals or 'undoings' can lead to a creative range of possibilities for learners to explore, while they experience freedom of choice within constraints.

SUMMARY

Mathematical topics arise as a collection of techniques and terms, but teaching needs to enable learners to enrich their experience by recognising important and pervasive mathematical themes.

In terms of the three threads—behaviour, emotion and awareness—a topic involves:

- language and techniques that the topic uses (behavioural thread);
- problems that give rise to the topic and to the contexts in which the topic may appear and which it resolves (emotional thread);
- images, connections and general awarenesses that feed and are fed by the topic (awareness thread).

These features are summarised in the structure-of-a-topic framework (Figure 2.1) which depicts the interweaving of three threads. Through training behaviour, learners can become competent but may, at best, reach instrumental understanding based on impoverished 'schema'. Through educating awareness, learners can develop relational understanding. The drive to learn comes from harnessing emotions.

3
MATHEMATICAL TASKS

The purpose of a task is to initiate mathematically fruitful activity that leads to a transformation in what learners are sensitised to notice and competent to carry out.

In learners' normal everyday activity, they initiate actions to satisfy their own motives, but educational activity is initiated in response to teachers' tasks. When a task is set, the *teacher's* intention is that the task will promote certain kinds of learning, but the *learner's* motive will depend on how the task is seen:

> ... even when students work on assigned tasks supported by carefully established educational contexts and by corresponding teacher-actions, learning as intended does not follow automatically from their activity on the tasks.
>
> Christiansen and Walther, 1986, p. 262.

The issue of different perceptions of the purposes of a task is explored in Section 3.1

Teachers set tasks because they believe that working on tasks will promote learning. But what learning might ensue from a given task, and how can tasks be selected to promote certain kinds of learning? A helpful way of addressing these issues is by viewing the purposes of tasks via different frameworks. This approach is discussed in Section 3.2.

There are not only the perceptions of the learners and the teacher to consider. Tasks also take place within a particular setting — the social and institutional environment of the classroom. This setting, which has been called the *milieu*, can have a significant impact on what learners do and how they respond. Section 3.3 considers the notion of 'milieu' in relation to the design of tasks.

The detail of a task has a crucial influence on what possibilities the task has for learners to access various experiences, and in what ways learners are constrained. Section 3.4 examines the consequences of different ways of presenting a task, as exemplified by a task based on the widely-used idea of 'arithmogons'. Then Section 3.5 considers the range of possibilities and opportunities provided by a specified task in a given situation.

Finally, Section 3.6 considers ways in which any task or set of exercises can be opened up so that learners are prompted to take the initiative and to make choices.

3.1 PURPOSES

Authors of tasks have purposes; the teachers who present the tasks to learners have intentions and expectations, and so do learners. But these purposes, intentions and expectations may not always be fully aligned! Consider the following process:

1. The author constructs a task or sequence of tasks.
2. The teacher decides to use or is requested to use the tasks.
3. The teacher imagines what the learners might do, what they might encounter (technical terms, techniques, facts, themes, powers, links to other topics, reference to other techniques and so on) and what they will practise (use of technical language, techniques and methods and so on).
4. The teacher considers what prior experience and encounters are necessary so that fruitful activity can be undertaken.
5. In the classroom, the teacher gives instructions.
6. Learners reconstruct for themselves what they think they have been asked to do, unconsciously altering the task so that it becomes something they think they can actually do.

Each of these steps in the evolution of a task provides opportunities for mismatches between intention and experience. Snyder (1970) coined the expression *hidden curriculum* to refer to aspects that teachers take for granted and do not teach explicitly. Some examples are:

- what it means to 'show your working' (one learner responded to the instruction 'show your working' by drawing a picture of someone at a desk evidently 'working'!);
- how textbooks are to be used;
- when learners should or should not contribute;
- the methods or contexts used to establish a technique but which are not referred to once the technique has been acquired.

The notion that there are different aspects to the curriculum has led to identifying the *intended* curriculum (TIMSS, 1997), the *implemented* curriculum and the *attained* curriculum (Schmidt, 1996). These distinctions apply not only to the whole curriculum but also to individual tasks. What is intended, what is activated (implemented) and what is attained or construed by the learners are often rather different.

AUTHORS' PURPOSES

If authors have gone to considerable lengths to publish a textbook, then presumably, so long as the learners do the tasks conscientiously, they will learn. Or will they?

The authors of a textbook are concerned that everyone using the book will know what to do. They want to minimise the effort required by the teacher to get learners active and doing mathematics. Consequently, they try to give precise instructions. The authors imagine that learners will, as a result of following these instructions, get a sense of some underlying structure or generality, and/or experience a technique sufficiently to internalise it, and/or come to appreciate and understand a topic.

However, it is not quite as simple as that. Chevallard (1985) pointed out that there is an inevitable shift as the expert-authors' awareness of the structure of a topic is transformed into instructions for the learner. He called this shift the *transposition didactique*, or didactic transposition. As Snyder's notion of the hidden curriculum suggests, there is a big step between learners doing things and learners actually learning what is intended. This is where teachers are necessary.

Teachers may be led to assume that the authors, as experts, have constructed tasks that will enable learning to occur, so the teacher's job is to make the tasks as accessible to the learners as possible. Accordingly, teachers set up some imagined context that they think will motivate the learners, then they demonstrate a technique on some examples, and finally they set learners to work on the authors' tasks. But this is rarely sufficient because it misses out the very heart of teaching—the interactions between teacher, learner and mathematics within the social milieu of the classroom. Such interactions are informed by the teacher's awareness of the underlying mathematical themes, the structure of the topic and the possibilities for learners to make use of their own powers.

No matter how authors try to make a task teacher-proof, the task can only be used effectively if the teacher is aware of its potential. But the situation is not simple.

On the one hand, if authors describe in detail the intentions, purposes and thinking behind each task, then the result will be a 'teacher guide' of huge size. Most teachers would not have the time to read and internalise such a guide, but even if they read it assiduously, they would still be likely to experience a version of didactic transposition. This is because, although the authors try to describe the awarenesses that lie behind the structure of the task, their words can easily be interpreted as instructions for the teacher. The spontaneity, the insight and the awareness that inform effective teaching can all be obscured by 'trying to remember to do all the things suggested'. This occurred when many teachers who tried to implement the National Numeracy Strategy in England without the benefit of an induction course found themselves 'doing as they were told' and so losing their creativity.

On the other hand, if authors do not describe the intentions and purposes behind tasks, then teachers have to decide for themselves whether a task is likely to be effective in their own context, or whether it needs modifying or augmenting in some way. The frameworks and suggestions described in this book are all intended to inform such decisions. Naturally enough, some teachers report that it is very difficult to adapt tasks in prepared schemes. This may be because they are not aware of the wider purposes of the tasks, or because they do not have time to experiment.

ATTITUDES AND BELIEFS

When learners come into a lesson in school, they have general aims of their own: to stay out of trouble, to attract attention, to show off to friends, to do as little as possible, to get through the lesson, to tackle a new challenge and so on. A major influence on what they will learn from a task is their attitude towards it: how far they wish to engage with the issues, as opposed to wanting to complete the task as quickly as possible.

Research by Dweck (1999) has shown how people's beliefs also affect their success in learning. Some learners act as though they believe that their intelligence and competence are fixed. Therefore they do not relish challenges but, rather, fear failure. When they get stuck they are likely to freeze or give up, and they take what they see as their failure personally. Others act as though their intelligence and competence can develop and mature. They tend to relish challenge. Encountering difficulty is grist to their mill. They do not even think in terms of failure, but rather in terms of opportunity.

Learners who are apparently successful may nevertheless baulk when the going gets difficult and may resist challenges because deep down they are fearful that someone will 'find out that they don't really understand'. By contrast, learners who appear to be unsuccessful can suddenly excel because they have been trying to achieve a deep understanding, while others have been satisfied with a superficial approach.

A belief that competence is limited can, perhaps paradoxically, be reinforced by constant and overt praising of success. This is because such praise sets up expectations which learners fear they cannot meet. Furthermore, if praise is over-used, learners soon recognise this and discount it. A teacher who wishes learners to see their competence not as fixed but as developing is likely to encourage them to replace 'I can't' with 'I can try, I can try harder, I can learn'.

TEACHERS' AND LEARNERS' PURPOSES

Brousseau (1997) has pointed out that there is an implicit 'contract' between teacher and learners, which he has called the *contrat didactique*. The teacher assigns tasks and learners do them: the result is intended to be learning. But this produces an inescapable tension:

> The more clearly the teacher indicates the overt behaviour expected from a given task, the easier it is for the learners to display that behaviour without actually generating it or being able to reconstruct it for themselves.
>
> Brousseau, 1997.

In other words, when instructions for a task are too directive, then it may be possible to carry them out without actually encountering the intended ideas behind the task.

Work on tasks can fail to produce intended learning in all sorts of ways. Learners' primary motive may simply be to do the task set by the teacher, rather than being concerned to learn something. Even when learners see their activity in terms of trying to learn, they may have considerable difficulties in understanding what it is that the teacher wants them to learn. This difference between how the learners and the teacher see the task is the major problem in trying to devise effective tasks. Moreover, learners often become absorbed in the *doing* of a task—carefully drawing lines, colouring shapes, cutting out and so on— they are being active, they are engaging in activity, but they may not be learning what is intended.

Nevertheless, learners may learn something important, yet what they learn may be different for each of them. Some learners may 'discover' or experience more sharply the advantage in working systematically, while others may experience the use of specialising in order to generalise for themselves. Still others may focus on the specific content and the techniques being used. Perhaps only a few will be aware of the central point as envisaged by the teacher.

Given these differences, reflection by the learners on what has been done and learned is essential. If the learners' motive is just to complete the task, the consequent activity may not be that intended by the teacher.

3.2   FRAMEWORKS FOR THINKING ABOUT TASKS

As a means for teachers to get a handle on the complexities of the purposes for which tasks are intended, researchers have produced several frameworks (discussed below) for thinking about tasks. Many teachers have found these useful reminders of the range of things learners need to be doing when working on tasks.

SEE–EXPERIENCE–MASTER

What is it reasonable to expect learners to achieve in a lesson? One learner might encounter a new idea or a new version of a more familiar idea for the first or second time. Another learner might gain further experience of something met previously. A third learner might be rehearsing some use of language or some technique in order to develop competence in it and so internalise it for later use.

These three possible outcomes give rise to a way of thinking about the purposes of a lesson, summarised by the framework *See–Experience–Master* (Floyd *et al.*, 1981). When learners first meet a topic, concept or technique, they are unlikely to make much sense of any explicit details: they will do no more than '*see* it go by'. Only with repeated exposure will they begin to get sufficient *experience* of it so that it becomes familiar and usable. With continued practice the learners can give less attention to the details and begin to *master* what the topic, concept or technique is about. Their growing competence and increased facility lead to what is often referred to as *mastery*. Mastery of a topic, concept or technique is intended to mean that it has been internalised, so that it functions at least semi-automatically, and leaves learners with a sense of 'having a good grasp'—they no longer need to focus their whole attention on the doing.

In the language of See–Experience–Master, the purpose of a task will include at least one of the following:

- exposing learners to something new (seeing);
- extending learners' experience of something that is becoming familiar (experiencing);
- practising behaviours to develop further competence and facility (mastering).

The See–Experience–Master framework can be effective without being made explicit to learners. If it becomes a guiding framework for mathematics lessons, then learners can form an image of habitual ways of working even though they may not know what to expect in a given lesson, or may not know precisely what it is that the teacher intends them to experience. That learners have such images of ways of working is an important element in effective teaching: it becomes part of the setting—the milieu of the classroom.

INNER AND OUTER TASKS

Although it is often advocated that learners be told what the purposes and goals of a lesson are, some of the teachers' purposes cannot be made explicit, at least to begin with, without being self-defeating. Tahta (1981) drew attention to an important but subtle distinction between what a task asks a learner to do explicitly (the *outer* task), and what the learner is expected to encounter while undertaking the task (the *inner* task).

The inner task refers to any mathematical powers that learners may find themselves using, as well as to the mathematical themes, topics, terms and techniques that they may encounter and the awarenesses that may be invoked. It is difficult to ensure that these aspects *will* be encountered, but it is likely that at least some learners will meet them and, through discussion, they can come to the attention of the whole class.

Tahta's notion of the inner task can usefully be expanded to include a variety of features:

- use of technical language, whether new to the topic or coming from previous topics, and arising from the need to express ideas;
- use of techniques previously encountered;
- use of mathematical powers such as specialising and generalising, conjecturing and convincing, imagining and expressing, and so on ;
- links to mathematical themes such as invariance amidst change, freedom and constraint, doing and undoing, multiplicity of representation and interpretation, and so on.

The term *meta-task* could be used to refer to far-reaching awarenesses that have implications or resonances beyond mathematics itself. Such awarenesses might include the knowledge that:

- many situations are susceptible to mathematical analysis;
- there are usually multiple ways of seeing and interpreting phenomena;
- there is some resonance between working mathematically with constraints and learners' struggles with institutional constraints;
- there is some resonance between extending and restricting mathematical meaning and the changes that occur in the interpretation of historical events.

There is no need for the teacher to be explicit about these resonances; indeed, mathematical work is more effective when the resonances are implicit.

A third manifestation of a task is the *personal* task. This involves recognising personal preferences and propensities, and then preventing them from blocking development. Such propensities might include a learner's tendency to dive into a task without standing back to consider how best to approach it, or a tendency to reach for a calculator without considering the significance of the numbers, or to collect examples and make tables of results rather than engage with underlying structure.

Being explicit about task purposes

These observations about the hidden features of a task lead to some apparent paradoxes. For example, if the aim of the teacher is to provide experiences from which learners can (re)construct for themselves some concepts, techniques or ways of thinking, then the more precisely the teacher articulates the lesson aims to learners, the more the outcome is likely to be in conflict with the aims. Furthermore, if the teacher describes an inner task to learners, it can make it more difficult for the learners to encounter that aspect freshly and to make it their own.

By thinking in terms of inner, outer, meta and personal aspects of tasks, the goals for lessons can be stated in a wide range of forms. The teacher could be precise about some technique ('Today we shall look at the naming of quadrilaterals, adding fractions, multiplying decimals, …'), or could mention some mathematical theme ('Today we are going to meet a mathematical theme we haven't seen for a few lessons …'). The teacher could even indicate that there are personal aspects on which learners might like to work ('Today there will be opportunities for you to find at least two different ways to approach a problem rather than just diving in …').

Of course, encountering powers, themes and propensities is one thing; learning from the experience is another. That is why teachers need to have sensitised themselves to these various aspects of tasks and to have strategies that will come to mind 'in the moment' when a choice is made to pause and pursue something that has arisen. Some of these strategies are discussed in Section 4.3.

MANIPULATING–GETTING-A-SENSE-OF–ARTICULATING

An important framework for thinking about tasks is summarised as Manipulating–Getting-a-sense-of–Articulating. The outer task involves learners in *manipulating* objects, which may be physical (including objects on an electronic screen), mental or symbolic, or, indeed, a mixture of all three. Meanwhile, the inner task and the purpose of the manipulation involve learners in *getting-a-sense-of* some underlying structure, pattern or relation-ship. As learners' sense of that structure gradually becomes more coherent, they will find that *articulating* (or depicting or symbolising) what they have understood becomes easier.

This process of transformation must take place regardless of what the teacher tries to make explicit. Even when learners are shown or told about a struc-ture, pattern or relationship, they must still internalise it for themselves. Put another way, just because they have been told something, it does not follow that they appreciate what has been said or can reconstruct it for themselves. However explicitly the teacher spells something out, learners must still trans-form it into their own knowledge.

This essential need for learners to reconstruct knowledge for themselves be-came the basis of a perspective known as *constructivism*. At first, construc-tivism was largely based on psychological literature and stemmed principally from the writings of Piaget, who himself built on numerous previous authors (Lerman, 1989; Davis, Maher and Noddings, 1990). A radical version of constructivism, developed by von Glasersfeld (1995), emphasises the mean-ings constructed by the learner. According to von Glasersfeld, learners construct meanings solely on the basis of their experience; he did not assume the existence of any 'objective' external reality to which the meanings are attached. Others (Cobb, 1994; Lerman, 1996) have preferred a social ac-count of constructivism, in which learners are seen as internalising *higher-psychological processes* (a term used by Vygotsky) displayed by those more expert than themselves. In this way, learners learn to speak, act and even think in ways that they pick up from the practices of those around them. Extreme forms of this social perspective claim that mathematical knowledge is not located in individuals but in language and other social practices. In each of these versions of constructivism, learners can be seen as going through a process of 'manipulating' leading to 'getting-a-sense-of', which gradually becomes easier to put into words, or 'articulate'.

USING AND REFINING MATHEMATICAL POWERS

How do learners deal with a new idea, gain experience of some technique they have met or develop their facility in techniques that are already familiar to them? The idea of 'practice'—getting learners to repeat similar computa-tions over and over again—has had only limited success in the many thou-sands of years that people have attempted to teach mathematics. A more informative approach is one that sees learners using and refining their natural or innate powers to achieve lesson goals.

Learners come to lessons with natural, or innate, powers:

- to imagine and detect patterns;
- to express those patterns in words, pictures, actions and/or symbols;
- to choose special cases of generalities in order to try to see what is going on;
- to re-generalise for themselves;
- to make conjectures;
- to modify those conjectures in order to try to convince themselves and others.

Whatever the topic, there is likely to be more actual learning when teachers provoke learners into using these powers than when they closely structure the work for the learners.

Of course, it is tempting for teachers to try to structure tasks so that 'anyone' can see what to do and can proceed unproblematically. But the more precisely the teacher structures the task, the less likely the learners are to get anything out of doing it. This is the *transposition didactique* of Chevellard, mentioned in Section 3.1. One of the effects of over-structuring tasks is to drain away all of the excitement and interest that was experienced by the designer of the task, so learners experience only a sequence of instructions. Once the teacher structures the task into subtasks, or helps by suggesting that learners make a table or generate some more examples or express a generality or justify their conjecture, the stimulus becomes a set of instructions. The original spark of interest may, indeed, be extinguished by worksheets.

Stein (1987) pointed out that an intention to 'teach thinking' can easily turn into 'teaching how to solve classes of problems', and this then involves teaching algorithms to learners so that they do not have to think! Since the whole point of a mathematical technique is to be able to solve problems without 'thinking from scratch', it might seem that learning techniques is an efficient way of learning to use mathematics. But techniques without thinking are essentially useless: they are inflexible trained behaviour—they are 'instrumental' rather than 'relational' in the sense of Skemp (1976) (see Section 1.1). If a problematic situation triggers a learner's awareness of a specific technique, it is easy for them to embark on that technique without thinking about whether it is appropriate or efficient.

Learners' pleasure in the spontaneous use of their powers is diminished to the extent that they are directed. A strategy that is likely to be more effective than pre-structuring a worksheet is to draw learners' attention after the event to the different ways in which they each used their powers, and to the ways in which such powers could be used more effectively in the future. When learners choose what to do, energy is created, creative moments are experienced and mathematical thinking comes alive.

3.3 MILIEU

Traditionally, mathematics education has been concerned with three aspects: mathematics, the learner and the teacher. In his comprehensive theory of teaching and learning, Brousseau (1997) augmented this triad with what he called the *milieu*—the learning and teaching environment.

The milieu includes:

- the classroom ethos and the characteristic ways of working;
- the degree to which learners and teachers are responsible for ensuring learning;
- the classroom organisation (social structure, resources and so on).

Each of these features influences how the teacher, individual learners and the class as a whole respond to a task. Bear in mind that a task is nothing more than an idea until a teacher and learners are working on it at a particular time and in a particular place, and until there is a milieu within which the task is embedded.

Some aspects of the milieu which can influence the pedagogic effectiveness of a task are:

- learners with no intention to learn will not respond in the same way as those who wish to learn;
- learners who are accustomed to using their powers to specialise and generalise are likely to engage in mathematical thinking;
- learners who are accustomed to having precise instructions for carrying out tasks are likely to have difficulty with tasks that require them to be creative;
- learners who are labelled as 'low attainers' or who are in a 'bottom set' may have low expectations of themselves and so may limit what they attempt and how hard they try;
- learners in a 'top set' may feel under constant pressure to perform or may be over-confident—either will colour their responses;
- learners in a room where they feel uncomfortable will find it more difficult to focus their attention;
- if the resources are inadequate, then the task as enacted will be different from the task intended.

Broadly speaking, the first two aspects are concerned with the ethos of the classroom and ways of working; the next three are concerned with where responsibility for learning resides and how learners' confidence and self-esteem are built up, thereby enabling learners to take greater responsibility; and the last two are concerned with the social or institutional setting as it is manifested in the classroom organisation.

The classroom ethos and ways of working, the degree to which learners rely on the teacher or take responsibility for themselves, and the classroom or-ganisation are all interdependent: they are not under the exclusive control of either the teacher or the learners. Rather, they develop as a result of inter-action, and they crucially influence activity in the classroom. These aspects of the milieu are now examined in turn, along with issues of learner confi-dence and self-esteem.

CLASSROOM ETHOS AND WAYS OF WORKING

Learners arrive at lessons with expectations, but these expectations are formed and changed by the classroom ethos and ways of working. For ex-ample, if learners have come to expect that every lesson starts with a new task, they may become dependent on the teacher always taking the initiative. Conversely, when learners' expectations are broadened, they are likely to be more open to changes in presentation and be prepared to respond to chal-lenges more readily. This more flexible approach is exemplified by the See–Experience–Master framework, in which learners expect to meet some new idea, to experience the idea repeatedly and to master that idea by developing competence and facility over time.

The teacher's role in creating expectations consonant with the effective learning of mathematics involves establishing ways of working that foster and sustain mathematical thinking. The aim is to provide conditions in which learners can educate their awareness as well as train their behaviour, since the two must go hand-in-hand if tasks are to be effective. Consequently, learners need to develop a stance in which problems are seen as challenges to work on and not as tests of memory, but at the same time they need to practise techniques so that these become automatic and internalised.

In shaping expectations, the classroom ethos is very important: an ethos that aims to help learners think mathematically must include a *conjecturing at-mosphere*. A conjecturing atmosphere gives rise to a way of working in which ideas are developed by encouraging learners to think out loud even when they are uncertain; the ideas are then mulled over and tested by the rest of the class. The whole class is engaged upon a collective enterprise in which everyone takes responsibility for making sense of what is said, rather than simply accepting or rejecting it—contributions are made in order to get help from others in clarifying or amending a conjecture. By stressing that statements are just conjectures, it is much easier to set them aside and go back to the main flow of the lesson—the conjectures can be recorded and returned to at another time if relevant. Most importantly, attention is diverted from being right or wrong, to improving and refining conjectures until they can be justified.

In a conjecturing atmosphere, anybody can be asked to explain their thinking so as to try to convince others. Thus, when two people disagree, each can try to persuade the other, and can try to offer examples that contradict the other's position, thereby initiating mathematical thinking. The point is not to be right, but mutually to reach for accord on what must be true. So when conjecturing and convincing are valued, then mathematical thinking is more likely to flourish. Conversely, when there is passive acceptance, reasoning and justification are discouraged.

An approach related to conjecturing has been advocated by Legrand. He stressed that it is important to introduce learners to what he called *scientific debate* (Legrand, 1993: see also Mason, 2001). Learners are seen as participants in a scientific community that develops by means of conjectures, modifications, proofs and refutations. This approach is based on ideas formulated by Lakatos (1976).

Legrand's guiding principles for scientific debate include:

- *Disturbance*: students must encounter and deal with conflict by means of ordinary rationality;
- *Inclusiveness*: every human being should have opportunities to understand the deep meanings of what is being taught;
- *Collectivity*: collective resolution of issues shows how to work with contradictions and how to respect the views of others.

Scientific debates can arise spontaneously when learners query a conjecture. When there are competing conjectures, they are often the result of different interpretations. If a teacher rides roughshod over these differences, declaring one or other conjecture to be correct without acknowledging or perhaps even being aware of the differences, then learners are likely to feel that mathematics does not make sense and that they 'just have to learn the rules'. The teacher may, therefore, need to end a discussion with two or three competing conjectures left unresolved. Such unresolved conjectures can be recorded on a conjecture board in the classroom and returned to at a later date, thus reinforcing the conjecturing atmosphere and perhaps providing a stimulus for independent work by learners.

Tasks that give rise to conjectures need not be elaborate. For example, the following task can promote conjecturing amongst younger learners:

> The teacher has made a two-dimensional cardboard shape, which she hides behind a screen. She gradually slides it up over the screen so the shape is revealed. As each new feature comes into view, learners are asked to conjecture what the shape could be: at first, the shape could be a kite, a rectangle, a pentagon, a right-angled triangle or something else (see Figure 3.1a); at the next stage, the possibilities are suddenly reduced (see Figure 3.1b).

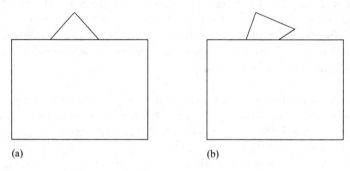

(a)   (b)

Figure 3.1 Cardboard shapes.

Of course, it is not sufficient for learners merely to formulate and discuss conjectures. To learn from doing a task usually requires learners to engage in reflection—to think back over what has happened. It is most important for learners to pay attention to *how* they sought to achieve an answer and which strategies were most effective. To this end, teachers need to emphasise to learners that the point of doing tasks is not just to get the correct answers. Stigler and Hiebert (1999) reported that Japanese teachers, rather than asking for a single answer, tended to ask 'In how many ways can you find ...', which shifts the focus away from 'a single right answer'.

Learners who shift from accepting what they are given to asserting conjectures are much more independent and able to cope with the unexpected. When encouraged to think mathematically, learners of all ability levels become able to tackle the unusual as well as the routine. They try to reconstruct ideas and techniques for themselves, as part of their reflection. Specific strategies for stimulating reflection are outlined in Chapter 5.

The classroom ethos and ways of working are often established within the first few days of a new school year or term, but can run into problems as time progresses. It is easy for practices to drift and turn into routines that no longer create the ethos originally intended. In addition, learners can become habituated to certain ways of working and may resent radical change. At this point, perhaps the most effective way for a teacher to make changes is to involve learners in trying out new ways of working, with a view to seeing what they think, while acknowledging that they may find it strange at first. This is to make the didactic contract open and explicit. Part of the contract might be to work some of the time in ways that are familiar to learners so that they have a degree of security, and some of the time in new ways.

RESPONSIBILITY FOR LEARNING

The relationship between the responsibilities of the teacher and of the learner is critical in the establishment of both a classroom ethos and the ways of working. The problem is really one of where the responsibility for learning lies—with the teacher or the learner.

Traditionally, emphasis has been on the teacher's responsibility. In this context, two widespread ideas are that the teacher needs to 'start where the learners are' and that learners need to be 'ready' to learn a particular topic. Each of these ideas has an important truth behind it, but can also be misleading. To ignore totally 'where the learners are' is to invite disaster—who would attempt to teach quantum mechanics to a four-year-old? But when the claim is that teachers need to know 'what the learners know' and to appreciate the learners' motivations, intentions, desires and propensities and so on, the request is absurd. What could a teacher do with all that information even if it could be gathered? With only one learner it would be difficult to use; with thirty or more it would be quite impossible.

Ever since Piaget's investigations into how children's minds work resulted in the defining of stages of psychological development (Piaget, 1950, 1977), the notion of *readiness* has pervaded mathematics education, with the focus being on whether learners are ready to learn a specified topic. There is certainly something in the notion of readiness. Learners are more likely to engage in a task that raises problems that they can appreciate or that creates surprise that they wish to resolve. Conversely, if a task involves some concepts or techniques unfamiliar to learners, they are going to find it difficult or impossible.

These notions of 'where learners are' and 'learner readiness' both imply that the teacher is the person who has to act on the relevant knowledge. In contrast, a central feature of a conjecturing atmosphere as a suitable milieu is that learners can take responsibility by using their initiative to make sense of mathematical ideas and to locate links between them. This process begins with the teacher asking learners to 'do things', but it can become more sophisticated as learners are encouraged or inveigled into becoming less dependent.

Learners can start to use their initiative when the teacher gets them to make choices. These choices may concern techniques, or mathematical objects to work on, or ways of seeing a problem or task. Thus, learners can be asked to choose or construct their own numbers or shapes, rather than have them provided by the teacher. For example, a task that starts with the instruction 'Choose a number, double it, add 3, multiply by 5, subtract 7, drop all but the units digit, ...' requires learners to choose their own starting number. The surprise which comes from the fact that everyone gets the same answer is an essential feature of the task and of how it can be developed into a significant piece of mathematics. Similarly, 'Draw a two-dimensional shape made up of straight lines, and calculate the total exterior angle' will lead to every learner getting the same answer. In such problems, the learners' answers are all related to their starting objects in the same way, and so the learners gain a sense of mathematical invariance. This is in contrast to a task in which learners work through a worksheet of given problems—the sense of invariance is missing because the starting points are the same for everyone.

It is common to have a set of exercises for learners to practise a specific technique; techniques can also be practised in tasks where learners have something to explore that involves constructing their own examples. For instance, rather than being given a list of numbers to factorise, learners could be given the task 'Find a number that has exactly seven factors'. This task would involve learners in a lot of factorising, but it has a purpose beyond that technique. As learners make up their own examples to test out generalities or to appreciate what the generality is saying, they are learning how to be mathematicians and, at the same time, are likely to enjoy using their own powers and initiative. The aim is to get learners to take the initiative and to treat mathematics as a constructive and creative activity.

The central issue of responsibility for learning comes to the fore when learners are uncertain or stuck. If they immediately put up their hands for attention, then they are reinforcing their dependency on the teacher. Instead, they can be encouraged to take responsibility for their own learning by developing some general strategies for dealing with difficulties. The strategies might involve simplifying the problem and trying some particular cases in order to see what is going on. Also, it can be helpful to locate the difficulty by asking 'What do I know?' and 'What do I want?', and perhaps writing down answers to these questions.

To encourage learners to take the initiative, some teachers have found it useful to have a poster labelled 'What to do when you're stuck!', which gives suggestions that can be pointed out. For example:

Ask yourself … ; ask a neighbour … ; only then ask the teacher.

Try some particular examples to see what is going on.

What do you want? What do you need? What do you want to find?

Mason, Burton and Stacey, 1982.

The teacher can use these poster questions regularly and repeatedly when learners ask for help. When such a practice has become usual, the teacher can ask a 'meta'-question, such as 'What do you think I am going to ask you?' Although at first this question may mystify some learners, it will help them become aware that the same sort of questions are being asked repeatedly. When teachers make their prompts less direct, learners tend to become less dependent. Moreover, as learners develop the habit of asking each other the poster questions, they will internalise them so as to be able to ask these questions of themselves.

What learners do and what choices they make influence the outcomes. By becoming aware that they have choices, learners may begin to challenge themselves.

LEARNER CONFIDENCE AND SELF-ESTEEM

Learners' personal confidence, self-esteem and expectations of themselves reflect to some extent the expectations of their teachers and parents. Learners assigned to a 'bottom set' often have, by that very allocation, reduced self-confidence and self-esteem. They expect less of themselves, and that can confirm the teacher's expectations. Thus a vicious cycle of descending motivation and expectation is set up.

Some teachers report that a more beneficial cycle can be established by expecting more rather than less—by trusting learners to take the initiative and by valuing learners' responses to the tasks set (Watson, 2002). Recall from Section 3.1 that Dweck (1999) has demonstrated that when learners adopt the attitude that 'intelligence is fixed and limited', then every time they meet a difficulty or a challenge, they assume they are at or near their limit. As a result they learn to avoid participating and so avoid failure, thereby reinforcing the descending spiral of low attainment. According to Dweck, when teachers avoid the vocabulary of failure and instead use a vocabulary of challenge and struggle, they can help learners shift away from a pattern of failure and can enable them to discover that they can use their powers effectively within mathematics.

If teachers consider that tasks involving mathematical thinking are suitable only for 'high attainers', then the result may be that low attainers are given a diet of routine and repetitive tasks on which they have already demonstrated their 'low attainment'. But if all learners are treated as possessing the powers necessary to think mathematically, and if those powers are evoked, developed and refined, the so-called 'low attainers' can transcend expectations.

Another factor that influences how much learners are responsible for their own learning is how talk takes place. When the exchanges are routinely between teacher and learners, the learners' dependency on the teacher is reinforced. One technique that a teacher can use to avoid the ping-pong of talk from teacher-to-learner-and-back is to get learners to talk in pairs. Giving learners the opportunity to discuss what they are thinking with someone else allows them to see if their ideas make sense before they put those ideas into the wider arena. Teachers who regularly get learners to talk in pairs report that learners are more likely to participate in subsequent whole-class discussions.

In fact, flexibility in the ways in which classroom talk is structured is likely to increase learners' motivation as well as decrease their dependence on the teacher. At the start of a task, it is useful for learners to do some initial thinking in order to make sense of the task for themselves and to formulate a plan of action, as well as to pinpoint anything that they do not yet understand. As a next step, a learner may pick up a better idea for an alternative approach by sharing ideas with other learners. A small group may work together effectively, but at some point the teacher may need to suggest a period of reflection to allow everyone in the group to reconstruct for themselves what has been said in the group. When teachers alternate periods of discussion and quiet work, it maximises the possibility that individuals will use the flow of ideas, not just to complete the immediate task, but to inform their future thinking. A task can be rounded off with reflection, which might include learners noting where the sharing of ideas was important.

How successfully a task is managed in the classroom depends on the relationship between teacher and learners. If there is trust and evident care, then a great deal can be achieved; if there is little trust and a feeling that control must be exerted, the issues of classroom management become intrusive.

CLASSROOM ORGANISATION

Among other things, classroom organisation includes the resources available for a task. A task can change fundamentally if one set of resources is substituted for another. For example, the task 'On a nine-pin geoboard, make as many different triangles as you can' becomes quite different if geoboards are unavailable. The ease with which triangles can be made, altered, demonstrated and compared on geoboards gives the task a momentum and excitement that is lost when learners can only draw triangles on paper.

Other aspects of classroom organisation are indicated by the kinds of questions teachers must ask themselves daily. In arranging the furniture, how can a balance be struck between the need for the interaction required for learning and that for maintaining order? Is it possible or desirable to have small or middle-sized groups? When is it useful for the class to move from a plenary discussion to working individually and then back to a plenary again? Are learners going to make posters and where will they be displayed? How will things be collected in at the end of the lesson? Will all learners do exactly the same tasks, in the same way, or is it possible to tailor tasks to individual learners? What will be the effect of giving some learners less demanding work?

To summarise, the milieu within which teaching takes place makes a vital contribution to what learning is possible. Yet the milieu is not entirely within the control of teacher or learner. Rather, it emerges through negotiation. This negotiation is more effective when it is clear to the learners that the teacher cares about both mathematics and the learners.

3.4   TASK PRESENTATION

When teachers present tasks to learners, there are dangers both in being too predictable and in being too innovative. If the teacher habitually uses only one method of presentation, learners can become so dependent that when they meet a different type of presentation, they are unable to cope. At the other extreme, using a different method every lesson may give the impression that there is no rhyme nor reason to how tasks arise, and this may put learners off. Most learners like both consistency and challenge. Maintaining an appropriate balance is a constant process of adaptation and experiment. The choices made by teachers will depend on their intentions and purposes, and on the timing of the relevant task.

This section takes a familiar task and considers nine different ways in which it could be presented or extended, before going on to consider dimensions-of-possible-variation for task presentation. Extending a task can be very useful as it enables learners to encounter the wider possibilities provided by the task: it can afford more scope for learners to use their innate powers, greater experience of important mathematical themes and/or more practice in order to support the internalisation of some types of computation.

PRESENTING AND EXTENDING TASKS

The different approaches offered here are all intended to make learners aware of the possibility of generalising, either in actions, words or symbols. Some of the approaches vary the mathematical problem presented to the learners, while others have different ways of organising the work in class. All of the approaches could be adapted for use with many other tasks.

A classic task known as *arithmogons* has been chosen as the illustration. It first appeared in McIntosh and Quadling (1975) and has been revitalised many times since; in particular, it was included in the National Numeracy Materials (DfEE, 1999a, b).

Because the focus in this section is on different ways of presenting a task, questions about the purpose of this task and where it fits in the curriculum will not be considered. However, the task is very versatile: versions of it can be used to generate practice in addition, subtraction, multiplication, division, differentiation, integration, finding the least common multiple and the greatest common factor, and more.

The original version of this task, as specified by McIntosh and Quadling stated (in relation to the figure reproduced here as Figure 3.2) that:

> There is only one simple rule to remember: *the number in the square must be the sum of the numbers in the circles on either side of it.*
>
> McIntosh and Quadling, 1975.

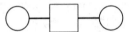

Figure 3.2   Circle and squares.

Various arrangements of circles and squares can be used, some requiring straightforward calculations, as in the one on the left in Figure 3.3, and others, more interestingly, requiring 'reversed' calculations, as on the right where the numbers in the squares are given.

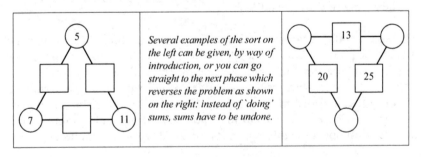

Several examples of the sort on the left can be given, by way of introduction, or you can go straight to the next phase which reverses the problem as shown on the right: instead of 'doing' sums, sums have to be undone.

Figure 3.3   Circles and squares (straightforward and reversed versions).

Presentation 1: Doing and undoing calculations

> Fill in the missing numbers, remembering that the number in the square must be the sum of the numbers in the circles on either side of it.

Almost any task can be presented as a sequence of calculations to be performed, rather like a set of exercises in a textbook. After working with some examples similar to that on the left of Figure 3.3, learners can be asked to make up their own examples and from these to create a version with numbers given in the squares. Neighbours can exchange these versions and try to 'undo' each other's. The point of working through examples is not only to practice the technique but also to obtain insight into the underlying structure.

The simple act of reversing a 'doing' question so as to produce an 'undoing' question can generate interest, if not surprise.

Presentation 2: Starting from a non-school context

Often it is possible to find something startling in a newspaper or on television or in some other situation outside school that will lead into the intended topic. Where that cannot be done, a story can be devised for an imagined context. With arithmogons, one story could be that, in an archaeological dig, tablets have been found on which some of the entries have been defaced. The learners can try to reconstruct what is missing using the underlying rules. This is an example of an artificial context that is faithful to the spirit of the work that archaeologists do. It also puts the focus on a central feature of arithmogons—the switch from *doing* to *undoing*. This switch occurs as you go from dealing with straightforward addition sums to working out what numbers you would have to add together in order to get the specified answers.

Views differ as to whether presenting a story around a task in this way helps learners. Some commentators think it makes mathematics more meaningful, while others take the view that learners can find an abstract topic intriguing to work on if care is taken about how it is introduced.

Presentation 3: Distributed work and pooling resources

Rather than the teacher providing several examples, a class could be divided into groups, each of which works on a different version of the task. This will generate a large number of examples for the whole class. For instance, in the arithmogon task each group of learners could use different starting values, first in the circles and then in the squares. These could be allocated by the teacher, or the groups could decide for themselves how to split up the work. Then, when all the groups have completed work, everyone will have access to several examples in order to look for what is the same and what is different about them.

After a period of working on the problem, learners can offer their ideas for making progress in a plenary discussion so that others can pick up on fruitful suggestions. Once learners have formulated their own questions, they can pool their ideas and decide which questions look tractable and worth pursuing.

Presentation 4: Starting with a hard or complex version

In this approach, learners are presented with a problem that is too difficult for them as it stands. They are then encouraged to make up their own related but simpler problems 'to see how they work'. As a result the learners participate more fully, being more interested in working on subtasks constructed by themselves rather than on the teacher's worksheets. In this case, the arithmogon task might take the form:

> Using the arithmogon rule, find numbers to fill the circles, which will give the correct answers in the boxes (Figure 3.4). Suggestion: before working on this, it might be sensible to try simpler numbers for the arithmogon on the left of Figure 3.4, and a simpler diagram for the one on the right!

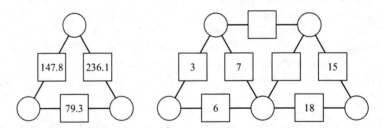

Figure 3.4   Circles and squares (using the arithmogon rule).

Presentation 5: What is the same and what is different about ...?

Two or three 'examples' are presented, and learners are asked to say what they think is the same and what is different about them. Then the learners are invited to construct examples of their own that differ in some way.

The arithmogon task might take the following form:

> What is the same, and what is different about the diagrams below (Figure 3.5)?
> Make up your own diagram that is different in some way.
> What are all the possibilities for varying the aspect that you have chosen?

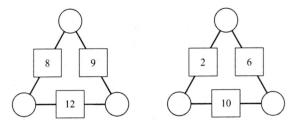

Figure 3.5 Circles and squares (same and different).

By varying the given numbers, attention is drawn to them as one dimension-of-possible-variation. An alternative way of opening up this dimension would be for the teacher to give numbers that force fractions or negative numbers or decimals to appear in the squares.

Another dimension-of-possible-variation could be opened up by varying the geometry of the diagram as in Figure 3.6 and again asking 'What is the same, and what is different?'

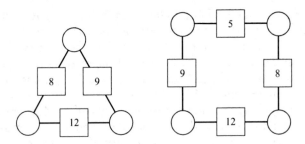

Figure 3.6 Circles and squares (varying geometry).

Other possibilities include providing diagrams with some numbers missing or diagrams with different operations.

Presentation 6: Starting in silence

The teacher draws the diagram shown in Figure 3.7 on a board or an over-head projector, and then fills in the squares in silence, pausing in a slightly exaggerated manner before writing each number, thus suggesting that the calculation is really being done. This continues for at least two or three examples. Learners can then be invited to come up and offer similar examples. The only comment made by the teacher, if any is needed, is to draw a happy or sad face beside the example offered, depending on whether it fits or does not fit what the teacher has in mind.

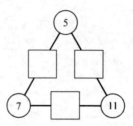

Figure 3.7 Circles and squares (starting in silence).

In terms of the mathematical themes of 'doing' and 'undoing', this presentation started with 'doing' in the form of addition. At some point during the silent start, the teacher might initiate the 'undoing' form of the task by putting up on the board an arithmogon with the squares filled in, and then writing question marks in the circles. Learners would then be expected to work on 'undoing'.

Presentation 7: Collective multiplicity

The teacher gives each learner a blank arithmogon triangle, and asks them to put numbers in the circles and then to calculate the corresponding entries for the squares according to the arithmogon rule. The learners are next asked to add the three circle numbers, and to add the three square numbers. The question 'What do you notice?' is likely to produce a response that the sum of the numbers in the squares is double the sum of the numbers in the circles. The fact that so many individual starting points all produce the same relationship is intended to raise the question of whether it is always so, thus enabling learners to experience the generality of a technique or procedure applied to many different examples.

Presentation 8: Starting with the general or with the particular

The teacher gives several examples and then poses a general problem. For instance:

> If someone rubs out the numbers in the circles and just leaves the numbers in the squares, can you always reconstruct the circle numbers?

Learners are then asked to try their own special cases in order to work out how to do this kind of question in general. More advanced learners may be able to use algebra. Learners can be asked to make up an 'easy' and a 'hard' question of this type (and do them), thereby revealing what they consider to be difficult, or perhaps discovering that there are no really difficult questions of this type.

Presentation 9: Using mental imagery

Instead of drawing a diagram, the teacher presents the task orally and asks learners not to draw anything. The teacher might suggest that learners close their eyes in order to cut out other distractions. For this approach to be effective, it helps to give instructions in the imperative:

> Imagine a triangle. At each vertex, there is a number. In the middle of each edge, there is a box with the sum of the numbers at either end of that edge ... .

Instead of starting with a triangle, the teacher could build up the image:

> Imagine a straight line (infinite in extent). Move it about in the plane: you can rotate it, or you can slide it about. Get a sense of all the possible choices (pause). Imagine a line-segment on your line, such that it has a starting and a finishing position. Slide the segment along the line, keeping it the same length. Now let it change in length, but still stay on the same line. Fix your segment, and then produce a second one. Make it the same length as the first. Attach one end of it to an end of the first segment ... .

This introduction is given in order to establish the freedom available, some of which will be constricted in what follows. Nevertheless, it provides a context of generality for the task, sketching out a dimension-of-possible-variation for later.

DIMENSIONS-OF-POSSIBLE-VARIATION FOR TASK PRESENTATION

The idea of dimensions-of-possible-variation was introduced in Chapter 1 to describe the process of varying the numbers and other features in a task while keeping the underlying structure unchanged. The same idea can be used to examine the range of possibilities that arise for the teacher in presenting a task.

The nine arithmogon examples just considered suggest the following dimensions-of-possible-variation in presenting a task:

- Who is active: the teacher, one or two learners, the whole class?
- How much do learners use their initiative?
- Does the task start with a simple version, with a complex version or with a generality to be specialised?
- Do learners work individually, in pairs, in small groups or collectively?
- What examples does the teacher use?

Each of these will now be examined in turn.

Who is active?

At the outset of a task, it may seem obvious that the teacher needs to be active, but this need not be the case. Teachers may start by directing learners to perform some actions rather than by demonstrating or explaining the task to the learners.

When learners are active, their actions may be physical (cutting or folding paper, using a finger to depict something imagined, or even using actual bodily movements as in 'people maths' with learners as objects). On the other hand, their actions may be purely mental as when they are asked to carry out a computation in their heads or to imagine the movements of points on circles.

The learner may be engaged in individual activity such as completing a worksheet, or in collective activity involving talk or showing. But learners' activity is not always productive. It may be mere 'busy work', or mindless activity where the mathematics has been left aside, or it may involve repeating things that they know how to do already. In other words, learners can be passively accepting even when they are apparently being active.

How much do learners use their initiative?

What does it mean for learners to 'use their initiative'? When they work on tasks from a worksheet or a textbook, there is usually no scope for them to use their initiative; nor is there any such scope when they are asked a question that has a single answer. However, learners do use their initiative when they have to make up their own tasks in order to explore a problem or they are asked to create examples or questions.

Having learners make up examples or questions is probably the easiest way for the teacher to get learners to use their initiative. They can be asked to devise an example or a question that will be useful to challenge people next year on the same topic, or to challenge the teacher or learners in another class. They can also be asked to construct a question which they think shows that they know how to do questions 'of the same type'. Other questions that encourage creative answers can arise when, after having been shown a 'doing' technique, learners are asked:

> If this is the answer, what could the question have been?
> Try to describe all the possible questions that are like this and that have this answer.
> Try to describe all the possible answers to questions of this type.

Another technique that the teacher can employ to get learners to use their initiative is to ask them to consider 'What is the same, and what is different?'

about two or three examples. (Of course, there is no scope for initiative if the teacher requires the learners to come up with 'what the teacher has in mind'.) Learners can then be asked to form their own examples and to do them to try to see what is going on. In each case, the discussion could be followed up by asking learners to make up further examples that share the features identified in the starting examples, and to decide what can be varied while still preserving these features.

In their book, *The Art of Problem Posing*, Brown and Walter (1983) recommended a strategy for generating questions—the *what-if-not strategy*. The teacher gets the class collectively to make a list of the features of a task, then picks one feature and asks 'What if that feature were to change? What could it become?', and so creates a list of possibilities. This is one way of getting learners to consider various dimensions-of-possible-variation.

Does the task start with a simple version, with a complex version or with a generality to be specialised?

It is tempting for the teacher to start with a simple example so that everyone in the class can see what is happening. But a constant diet of simplicity leads learners to depend on the teacher always starting in the same way. By sometimes starting with a complex example, the teacher can help learners to gain a different sense of where they are heading. Furthermore, learners can participate in the process of simplification or specialisation, thus using their own powers rather than depending upon the text or the teacher to do it for them.

Sometimes a generalisation is made easier by starting with a complex occurrence. For example, when teaching methods for factorising quadratic expressions, rather than give a simple expression to factorise, the teacher could present the expansion

$$(3x + 2)(5x - 1) = 15x^2 + 7x - 2$$

and ask 'How can you spot the factors if you are given only the expanded product?' The learners could then make conjectures and devise simpler examples to test those conjectures.

Do learners work individually, in pairs, in small groups or collectively?

The scope for variations here is huge. For instance, at the start of a lesson, individuals could be asked to do something and then the results could be shared. Perhaps each learner could create their own example so that the class as a whole quickly generates a pool of different possibilities. This may lead to a surprise about what all the answers have in common.

At other times it makes sense to start a lesson with learners in pairs, with each pair taking a different role. Alternatively, the teacher can get small groups doing apparently different things which nevertheless all end up with the same idea. Another possibility is that everyone can be asked to do the same thing so that they all have the same experience—and this will also provide a check on accuracy.

The various ways of working have different strengths:

- *Individual* work allows learners to review, consolidate and develop their facility, as well as to reconstruct for themselves.
- Work in *pairs* allows learners to try out ideas on each other before offering them to a wider group; it also provides an opportunity for learners to consider something that has happened or been said, and to generate more ideas about this than an individual is likely to produce when working alone.
- Work in *small groups* allows a multitude of ideas to be generated, and also allows a large task to be split up amongst several people; with discipline, small groups can provide a forum for discussing ideas, modifying conjectures and coming to a consensus with supporting reasons and justification.
- *Collective* and *plenary* work allows everyone to hear about novel ideas and approaches, and to see teachers or peers displaying their own mathematical thinking. However, it is easy for individuals to be caught up in collective actions without internalising the ideas behind the activity, and so periods of individual work are also needed.

What examples does the teacher use?

An important feature of task presentation is the type and level of difficulty of the examples chosen by the teacher. These choices are mainly determined by the teacher's awareness of which dimensions of the examples will need to be varied in order for learners to see the permissible range of change. Learners too can develop the habit of seeking out dimensions-of-possible-variation, testing them with examples, and locating relevant ranges-of-permissible-change.

3.5   POSSIBILITIES AND OPPORTUNITIES

The most important feature of a task is that it contains some challenge without being overly taxing. That is, it neither appears to be trivial nor does it appear to be beyond the learners' capabilities. Tasks which demonstrate to learners that they are capable of more than they imagine can do wonders for building their confidence; on the other hand, tasks that have been 'suitably simplified' to be well within the learners' capabilities are likely to be seen by them as not worthwhile, and so may demotivate rather than inspire.

A great deal depends on the atmosphere of trust between the learners and the teacher. This includes trusting that the teacher will set tasks that will be rewarding and engaging, even if at first they seem unpromising. It also encompasses the teacher's belief that learners can choose responsibly and that their responses are worth taking seriously.

If the didactic contract (see Section 3.3 above) is exploited by the teacher, then learners can develop a sense of identity through the growth of their confidence and through being surprised at themselves. This requires the teacher constantly to induce learners to work for understanding rather than just to follow explicit instructions.

Different tasks may offer different possibilities for learners to:

- encounter specific mathematical ideas and techniques;
- experience the use of their mathematical powers;
- enrich their awareness of mathematical themes;
- become aware of and work against their personal propensities (propensities such as diving into a task without considering the best approach);
- gain facility and competence in the use of a technique.

However, a possibility is of little use if the learner or the teacher is not aware of it as an opportunity at the moment when it matters. Therefore it is useful to distinguish between *possibilities* provided by a particular task and *opportunities* recognised and seized by teachers and learners: a task may have possibilities, but these may not be recognised as opportunities to be exploited.

An observer of a lesson always notices some possibilities the teacher did not act upon. The teacher may have decided not to follow up some of these or may have been unaware of them. Whatever possibilities a task might have, it is what the teacher and the learners are aware of at the time that matters. Hence, it is not the task alone that provides opportunities, but rather the overall situation; the possibilities provided by a given task or type of task, the established ways of working in the classroom and the propensities of the teacher and the learners all influence what opportunities are available. Also bear in mind that, because the format of a task shapes the activity that learners engage in, the format affects the opportunities available to them; slight variations in how a task is specified can alter what learners are likely to do with it.

It is one thing for teachers to imagine that they will listen to what learners have to say, and will then let the lesson develop in response to the learners' needs, but quite another to do this in practice, because of the various forces acting in the classroom and also because of the 'tunnel vision' that develops in the midst of a lesson. Equally, once learners have begun to use their initiative to explore around a topic and to make up their own examples, the structure planned for the task may be over-ridden. Although an observer may be aware of all sorts of missed opportunities, these choices may just slip unnoticed past the teacher. Indeed, to be aware of the missed opportunities, the teacher usually needs a colleague or other observer in the lesson to point them out afterwards.

In any situation where a task is being used, there will be many different possibilities, which learners may or may not be aware of, but which the teacher may be able to make into opportunities. Recognising these possibilities may not always be easy. To help, various questions are set out below, grouped under the headings of behaviour, emotion and awareness, respectively. Each main question (in italics) concerns one set of possibilities, and is followed by a set of subsidiary questions that examine the aspects relevant to the heading. For each set of possibilities, only the features relevant to the heading are mentioned, although the possibilities may contribute to other headings.

Throughout this section we use the term 'task-situation' as a reminder that each task is set in a particular situation, and the possibilities under discussion are a product of both the task and the situation in which it is set.

BEHAVIOUR

The following questions principally concern behaviour. They involve possibilities provided by the task-situation for learners to:

- manipulate objects and so experience or internalise something;
- see new ideas go by, experience more familiar ideas and gain prowess in the use of techniques and relevant vocabulary;
- produce something worthwhile.

What possibilities are provided by the task-situation for learners to manipulate objects and so experience or internalise something?

Being able to manipulate objects, whether physical, mental, diagrammatic, symbolic or on a computer screen, can appeal to learners' playfulness and give them confidence. However, it does not necessarily advance their mathematical thinking, and so tasks need to incorporate ways of helping learners to go on to acquire a growing appreciation of the mathematics of the situation.

Related questions that may help teachers to focus in more detail on aspects of the learners' experience of manipulation include:

- Does the task-situation have possibilities for learners to manipulate objects?
- Are there possibilities for learners to do things physically, mentally, diagrammatically, symbolically or on a computer screen?
- Are learners asked to articulate their thinking?

- Are there possibilities for learners to use a range of powers and to encounter one or more mathematical themes?
- Are there possibilities for 'standing back' and reflecting on the activity, and for seeing where the task fits in a wider view of the topic?

A key aspect of gaining confidence in an idea and internalising that idea involves developing the vocabulary to articulate it and the techniques to apply it to tasks.

What possibilities are provided by the task-situation for learners to see new ideas go by, experience more familiar ideas and gain prowess in the use of techniques and relevant vocabulary?

The framework underlying this question is See–Experience–Master. Techniques can be consolidated, for instance, by learners constructing their own examples, or while learners are exploring an area of mathematics that is unfamiliar but which calls upon the technique.

Related questions that may help teachers to focus in more detail on aspects of the learners' experience of going from seeing to mastery include:

- Does the task-situation involve the use of familiar objects, techniques and vocabulary, with a view to encountering, discovering or reconstructing something less familiar?
- Are possibilities provided for learners to move from struggling to say what they are seeing or thinking, and progress to refining their articulation?

Tasks that attract attention to the new while exercising the familiar are multi-layered. They are much more likely to be engaging and to yield a sense of satisfaction than tasks that concentrate entirely on practising some technique.

What possibilities are provided by the task-situation for learners to produce something worthwhile?

Having an overt product of some sort can focus learners' attention on the task. There is a wide variety of such possible products. For some learners, the most authentic product is one that will change their lives; this might involve, say, reorganising aspects of school life (see The Open University, 1980; Mellin-Olsen, 1987; Frankenstein, 1989; Skvovemose, 1994). Others want tasks to be based on things that adults do at work or at leisure. More modest products are a written report, a poster for the classroom or a web page.

Related questions that may help the teacher to focus in more detail on aspects of the learners' experience of producing something include:

- Does the task have a tangible product—be it a proposal for a new way of organising something in the learners' lives, a physical model, a poster or a computer design?

One danger of concentrating on a product is that it may divert attention from the deeper mathematical content. Instead, some curriculum projects prefer to 'make things real' for learners by engaging their powers, especially their power to form mental images. (This is the basis of the Realistic Mathematics Project of the Freudenthal Institute; see Gravemeijer, 1994.)

EMOTION

The following questions principally concern emotion. They deal with possibilities in the task-situation for learners to:

- participate in making choices and take initiative;
- experience surprise as a motivation for developing a mathematical explanation for a phenomenon;
- sustain their initial enthusiasm long enough to learn from the experience and so gain satisfaction.

What possibilities are provided by the task-situation for learners to participate in making choices and take initiative?

The more that learners are involved in making choices, the more likely they are to see mathematics as a constructive enterprise and not as a collection of abstract techniques to master. Moreover, as learners come to understand how to open up tasks to exploration and inquiry, they are likely to increase their commitment, take initiative and get more enjoyment from what they do.

Related questions that may help the teacher to focus on aspects of the learners' experience of participation and taking initiative include:

-   Can learners choose which examples they use in order to see what is going on in some generality?
-   Can learners choose which approach to use for a given problem?
-   Can learners decide what problems to tackle?
-   Can one learner decide what problems will be set to another learner?
-   Do learners have a choice of task?
-   Are learners given choices between easy, hard or challenging problems of a given type?

Learners who are offered the chance to choose will need help to learn from the experience through focused reflection—they are 'learning how to learn'. In this way, they are likely to become more engaged and less dependent on the teacher for routine matters, thereby enabling the teacher to focus attention on the more sophisticated aspects of the topic being taught.

What possibilities are provided by the task-situation for learners to experience surprise as a motivation for developing a mathematical explanation for a phenomenon?

Surprise is an important feature of tasks because it creates the energy to engage with the mathematics—learners are more likely to engage with a task that surprises them and that they wish to resolve.

Related questions that may help the teacher to focus in more detail on aspects of the learners' experience of surprise include:

-   Do tasks have the potential to generate surprise?
-   Are learners likely to become sufficiently intrigued to want to explain something that they initially find surprising?
-   Is there support for learners to move from their initial surprise to a recognition that the feature is, in fact, typical of the situation, and then for the learners to go on to further generalisation?

Although surprise is a useful starting point, it is crucial to move beyond it to engage in sustained activity.

What possibilities are provided by the task-situation for learners to sustain their initial enthusiasm long enough to learn from the experience and so gain satisfaction?

Learner enthusiasm is an important ingredient in learning; among other things, it encourages the sustained work needed for progress.

Related questions that may help the teacher to focus in more detail on aspects of the learners' experience of sustained work include:

- Is the initial task likely to sustain interest?
- Are learners likely to want to find out what initially disparate-looking objects, techniques or results have in common?
- Are there possibilities to generalise?
- Is there some context that will spur learners to draw their thinking together and make sense of it?
- Do learners respond positively to a competitive aspect?

Satisfaction can come from a sense of using and developing one's personal powers, and coming to understand something or to integrate ideas that previously seemed disparate. Satisfaction may also come from a learner's sense that the task is relevant to life outside school or to reorganising some aspect of school life.

AWARENESS

The questions here mainly concern awareness. They involve possibilities provided by the task-situation for learners to:

- develop and refine their powers;
- experience various types of mathematical questions;
- discuss and try out ideas with others;
- reconstruct ideas for themselves.

What possibilities are provided by the task-situation for learners to develop and refine their powers?

This question addresses the possibilities for learners to engage in doing things, not only physically but also emotionally and mentally. Moreover, it encompasses possibilities for learners to pause in the doing, draw back, and try to make sense of the activity. It also covers possibilities to move from particulars to generalisations, and to construct special cases that fit or exemplify some expressed generality.

Related questions that may help the teacher to focus in more detail on aspects of the learners' use of their powers include:

- What mathematical thinking processes are learners expected to use spontaneously and what powers might have to be triggered?
- Are learners expected, without prompting, to draw up a table, or draw a graph, or make up their own examples, in order to formulate and test a conjecture?
- Are learners being trained to work more and more independently?

There is considerable variation in what learners are expected to bring to the task-situation and to engage in as part of the activity. Tasks can be seen as vehicles for training learners to be increasingly independent as mathematical thinkers, and this makes the learning, and hence the teaching, increasingly efficient over time.

What possibilities are provided by the task-situation for learners to experience various types of mathematical questions?

A range of such possibilities is important because the types of questions that learners are asked will influence their view of mathematics.

Related questions that may help the teacher to focus in more detail on aspects of the learners' experience of mathematical questions include:

- Do learners experience a wide range of types of questions and questioning?
- Do some of these questions call upon learners to construct or reconstruct mathematical objects for themselves?

On the one hand, learners may form the impression that mathematics is a creative and constructive enterprise, involving the exercise of human powers; on the other hand, they may come to believe that mathematics is a closed, narrow, routine collection of techniques to be applied to stylised and unrealistic questions. Watson and Mason (1998) explored the variety of types of questions that mathematicians ask, and showed how standard and routine tasks can be adapted to display this variety.

What possibilities are provided by the task-situation for learners to discuss and try out ideas with others?

Understanding does not arrive suddenly as the result of work on some task. It matures over time. Many educators have observed that the gradual process of developing ideas and competence is enhanced by discussion.

Related questions that may help the teacher to focus in more detail on aspects of the learners' experience of discussion include:

- Are learners asked to do things and then to talk about what they have done?
- Do learners talk to each other in order to clarify their ideas?
- Are learners able to listen to others' comments and ideas, and then respond to them?

Only when learners can talk reasonably fluently about the techniques or ideas involved is it useful for them to record these things.

What possibilities are provided by the task-situation for learners to reconstruct ideas for themselves?

If learners are always required to follow instructions, and especially if those instructions have been refined so as to be perfectly clear and unproblematic, then learners are likely to respond by following the instructions with little thought. Therefore it is essential to provide scope for learners to reconstruct ideas on their own.

Related questions that may help the teacher to focus in more detail on aspects of the learners' experience of reconstruction include:

- Are learners encouraged to reconstruct facts and techniques for themselves in novel situations or contexts?
- Do the tasks that are set require learners to use only memorised techniques in the form in which they were given?

Bear in mind that education occurs when learners do things mindfully so that they can learn from the task. This makes them more likely to be flexible when meeting a fresh situation and ready to adapt to the new setting.

3.6 OPENING UP TASKS

Searching for an ideal task that would be 'just right for my situation' is neither necessary nor, ultimately, fruitful. It is much more effective for a teacher to develop ways of creating tasks in the classroom and of adapting already existing tasks. Such techniques fall under the general heading of 'opening up tasks' as they frequently make a task richer than its original version. This section briefly summarises a number of techniques for opening up tasks.

WARM-UP TASKS

One purpose of warm-up tasks at the beginning of a lesson is to provide a mental version of callisthenics. Another purpose is to refresh learners' memories of recent work and, in particular, to allow learners to practise techniques—both those that are not quite internalised and those that are. Other purposes are for learners to use their innate mathematical powers and also to encounter important mathematical themes.

Starting a lesson with an opportunity to detect and express a general pattern is a useful strategy as it exercises learners' natural ability to imagine and to express generality; alternatively, learners could be asked to specialise. A set of generalising and specialising tasks could be based on a question such as:

Which is bigger, 57 or 75?

and might take the following form:

Generalising	Specialising
How do you tell which of two numbers with two digits is the larger? (Could be more digits, could be decimals or fractions, as appropriate.)	Write down a number between 10 and 100. Now write down a number that is smaller; then a number that is at least 23 smaller; then at least 23 greater; ... . (Adapt to other types of number as appropriate.)

Another pair of generalising and specialising tasks is:

Generalising	Specialising
What do all the numbers in a row have in common? 2, 20, 200, 2000, ... 3, 6, 9, 12, ... 5, 10, 15, 105, ...	Write down two numbers that are both odd, and which are larger than, say, 100, and which you think no-one else in the room will write down. Explain your choice. (Other properties could be, for instance, that both numbers are odd squares, or divisible by, say, 13, and so on.)

More generally, any mental or oral warm-up task can be adapted so that learners' powers, particularly of generalising and specialising, and of imagining and expressing, are exercised.

CREATING TASKS

One previously mentioned strategy for creating tasks is that advocated by Brown and Walter (1983): the what-if-not strategy. In this, learners are asked to choose an attribute of a given problem or situation and to consider 'What if it were not that?' For example, for the Fibonacci sequence 1, 1, 2, 3, 5, 8, 13, ..., some attributes are: it starts with two numbers; the two numbers are both 1; and each number is obtained by adding the two previous numbers. One of these attributes is picked, and a new task is created by varying some feature of that attribute. Take the last attribute given above for the Fibonacci sequence: if it were not 'adding', it could be 'multiplying' or 'dividing'; if it were not 'two previous numbers', it could be three or four or more. Thus new problems—and tasks—are created. This is a way of exploiting dimensions-of-possible-variation.

Prestage and Perks (1992) did something similar by looking at the unstated features of problems. As an illustration of their approach, consider the problem:

> How many 7 cm strips of paper can be cut from a paper strip that is 161 cm long?

This can become a host of different tasks when the (unmentioned) width of the paper strip is specified. What if the strip were 7 cm wide, or 14 cm wide? What if the cuts were on a diagonal?

A more general version of these techniques for creating tasks involves presenting learners with an object to look at—a diagram, a poster, an animation, a formula or even a collection of exercises—and then asking them to 'say what they see'. They may make distinctions, observe relationships, locate properties, characterise objects according to those properties and so on. What is 'seen' usually generates further tasks to explain whether 'it always works' and why it might 'work'. In addition, when someone sees something not seen by others, further elaboration is necessary, often leading to yet more 'tasks'. This type of task can serve as a model for how learners themselves may think when confronted with a new situation: they may ask themselves what they actually see (as distinct from what they imagine to be there). In essence, what they are doing is seeking invariance in the midst of change.

Becker and Shimada (1997) examined differences between tasks used in Japanese and American classrooms, and subsequently made suggestions for devising tasks. They suggested taking a physical situation involving some variable quantities in which mathematical relations can be observed and asking learners 'What relationships can you find?' The aim is to find relationships that transcend a particular instance, so follow-up questions might include 'What can you change and still have a similar relationship?' In the case of a plastic cup rolling on the floor in a circle, the question might be 'How does the size of the circle depend on the dimensions of the cup?' More generally, the Manipulating–Getting-a-sense-of–Articulating framework (Section 3.2 above) can serve as a reminder of what is being looked for during work on such a task.

SUSTAINING WORK ON TASKS

Even when tasks promote doing, seeing, talking and recording, there is still the problem of how teachers can best encourage learners to adopt appropriate motives for their activity. For example, how can teachers get learners to concentrate on developing generalisations rather than on simply identifying a particular pattern? Tasks which embody the essential requirements for learning to take place also need to ensure that learners have opportunities to reflect on their learning. Although there are no infallible methods for ensuring that the focus of learners' work becomes their own learning, teachers have devised various kinds of tasks that are more likely to confront learners with the need to be aware of their learning.

One such type of task is a *game* in which learners are called upon to see and express certain patterns or connections. A carefully constructed game can be a useful way to introduce a topic or concept if, in order to play the game effectively, learners need to use that topic or concept. The pleasure that learners get from playing games serves to engage them in situations where they can practise the use of technical language when describing what they are doing or seeing.

An example is provided by the 'Game of 31', in which players alternately choose a number from 1 to 5 and calculate the cumulative total. The first player to reach a target of 31 is the winner. This game was studied in detail by Brousseau and colleagues during the 1970s in the development of their sophisticated theory of didactic situations (Brousseau, 1997).

Playing the Game of 31 can:

- lead to a desire to find a winning strategy;
- open up the possibility of working backwards, which is a useful mathematical strategy;
- provide a lead into the topic of remainders;
- offer an opportunity to generalise the dimensions of 'choose a number from 1 to 5' and of ' a target of 31', with a view to formulating a general strategy.

Another relevant type of task involves situations that appear to give rise to *contradictions* or *surprises*. In these tasks, learners need to sort out what is happening, resolve differences of opinion or conflicting explanations, and find some way to account for what is going on. Learners are called upon to explain things to each other and to locate differences and agreements in their explanations.

An example of such a task is:

> Consider a small triangle and a large triangle with the same angles. Which triangle will have the larger angle sum?

Another example is:

> Write down two different numbers whose sum is 1. Which will be larger: the square of the larger plus the smaller, or the square of the smaller plus the larger?

In general, these situations result in learners making explicit their own difficulties. Discussion is often used to do this, and various strategies can be employed to promote discussion:

- Learners devise their own questions and give these to others not only for them to do but also to discuss whether those questions are easy, hard, similar to one another and so on.
- Learners are encouraged to construct stories that incorporate a specified problem.
- The teacher sets up a 'conflict-discussion lesson' in which the problems to be worked on are ones about which learners are likely to disagree but be convinced of their own correctness. In this way, conflicts are built into the lesson and are resolved by groups of learners discussing their different results.
- The teacher builds reflection into the end of an activity by asking the learners to present verbal or written accounts of what they have learned.

These strategies help learners become aware of the purpose of the task, so they approach it with appropriate motivation. One problem for teachers in trying to use such strategies is that the activity they produce often runs counter to the teacher's views on the proper way to encourage learning. Thus, in many tasks, explaining the mathematics involved—as distinct from clarifying the task—is likely to defeat the purpose of the task. Bell commented:

> For many teachers it is difficult to organise discussion so that misconceptions are brought to the surface. In fact, some of the teachers organised their learners by groups, but then spent their time touring the class

helping the groups in the usual way, without any particular emphasis on exposing and discussing the misconceptions.

Bell, 1987, p. 53.

PRACTISING TO PERFECTION

There are some learned skills that remain stable over very long periods, such as riding a bicycle and playing a piano. However, both understanding a concept and being fluent in a technique tend to fade if not used. So, it is not a matter of learners practising until a technique has been integrated into their functioning, and then never forgetting it (Li, 1999). Rather, techniques, words, phrases and concepts need to be revisited every so often in order for them to remain fresh. Fortunately, learning mathematics is not like memorising long passages of text, which, because they are unrelated to other texts, need to be refreshed as a whole. In mathematics, tasks can be structured so that the old is constantly being used in order to encounter the new. The See–Experience–Master framework (see Section 3.2 above) serves as a reminder of this.

Learning effectively from practice can be encouraged by tasks involving either *challenges* or *explorations*; such tasks can also broaden the ethos of the classroom.

Challenges can be offered so that learners' attention is focused on discovery, but in the process learners are called upon to exercise particular skills to a sufficient extent that the skills become automated. An example of this is provided by a task originally devised by the Madison Project (Davis, 1966), which asks learners to fit the same number into each box in a quadratic expression such as

$$\left(\square \times \square\right) - \left(5 \times \square\right) + 6 = 0$$

and to see if they get a true statement. Some numbers work and others do not. Are there any rules for deciding which numbers will give the true statements?

Because learners are set the task of looking for such rules, this may be how they view what they are engaged upon. But the teacher's intention is to get the learners to practise substituting values for a variable and to gain an awareness of what a variable is—the actual rules themselves are seen as unimportant for future mathematical learning.

Other tasks with similar purposes are based on board games that require learners to count. These provide practice in counting while attention is focused on the game.

Practice can also be gained through *exploration*, not least because exploration in one topic often calls upon techniques and language from other topics. Two examples of exploratory tasks are:

- Create a set of paper circles showing the sizes (to scale) of the Sun and the nine planets of the Solar System. Construct a scale drawing of the Solar System, showing the distances of the planets from the Sun.
- Compare the amounts of food eaten by different people and different animals, in comparison with their body weight (Sierpinska, 1994).

Here, learners are free to choose what they will consider, and how to go about it. Both tasks involve practice in calculating and scaling.

Other exploratory tasks might involve learners in constructing their own examples and then generalising. Such tasks are another way of ensuring that learners' practise, without them thinking that they are 'simply doing routine exercises'. Hewitt (1996) recommended the use of such tasks, because the learners' attention is at least partly on the example, while the technique to be learned is in the background. In this way, the details of the technique are likely to be integrated into the learners' thinking. Consequently, learners become competent at the technique, and then require a minimum of attention to carry it out. Sets of routine exercises rarely achieve this because attention is directed towards the very technique to be automated.

ROUNDING OFF A TOPIC

Tasks that help learners summarise and appreciate the key features of a topic are a useful way of rounding off a section of work. One relevant type of task involves getting learners to invent examples of the kinds of questions that they have been working on. They could be asked to produce:

- a question that is 'of this type' but is really easy (what makes it 'of this type'?);
- a question that is 'of this type' but is really hard (what makes it 'really hard'?);
- a general question 'of this type', or perhaps several questions that vary different dimensions-of-possible-variation;
- a question that demonstrates the ability to do questions 'of this type';
- a question that would be a challenging problem for learners in another class or next year.

This sort of task draws learners' attention away from the doing of specific tasks and into the scope of the tasks themselves, as well as into the general class of problems they can now tackle. They reveal learners' awareness of dimensions-of-possible-variation and associated ranges-of-permissible-change that constitute the 'task type'. They also support learners in becoming aware of 'types of task' so that the learners are more likely to recognise a particular type of task when they meet it again later.

Other types of task that promote thinking about a topic are:

- brain-storming sessions which invite learners to think about the three threads of a topic—behaviour, emotion and awareness;
- learners constructing their own web diagram for a topic.

In general, each task has both mathematical and pedagogical dimensions-of-possible-variation. It is vital for learners to appreciate the mathematical dimensions. Their interest and involvement can be maintained by exploiting the pedagogical dimensions.

SUMMARY

In this chapter, attention has been drawn to the different perspectives held by authors, teachers and learners concerning tasks.

Some frameworks for thinking about the design and use of tasks have been introduced: the *See–Experience–Master* framework, the distinction between *inner* and *outer* tasks, and the *Manipulating–Getting-a-sense-of–Articulating* framework. The former framework views learners as being on a trajectory in which they see some things go by, gain further experience of things encountered previously and then develop competent mastery over the topics, ideas or techniques they have met. According to the Manipulating–Getting-a-sense-of–Articulating framework, learners will find that their ideas become clearer and make more sense when they manipulate objects (physical, mental, symbolic or virtual) with a view to trying to get a sense of what is going on, and of being able to articulate that sense succinctly. All this happens most effectively when learners are using and refining their own powers.

Overall, the frameworks provide a richer and more integrated sense of what any one lesson or task may be contributing to learning.

Attention has also been drawn to the importance of the *milieu*—the ethos, ways of working and social situation in the classroom. The more independent the learners can be, the more likely they are to be engaged. To this end, the notion of a *conjecturing atmosphere*, in which learners participate in scientific debate, has been introduced.

The idea of *dimensions-of-possible-variation* has been extended to the ways in which a task is presented. This was illustrated by nine different ways of presenting the same task. It was also suggested that too much variety might be upsetting, but too little variety produces dependency. The notion of dimensions-of-possible-variation has been used to formulate questions about the nature of learner engagement with tasks, and this led into the distinction between the *possibilities* in a task, and the *opportunities* in the tasks that are actually adopted by learners and teachers. Possibilities were related to behaviour, emotion and awareness as aspects of the psyche.

Finally, proposals were made that any task, no matter how apparently routine, can be opened up to challenge or exploration in order to keep learners usefully engaged. It was suggested that practice is more efficient when learners are engaged in some larger task where they are making up examples on which to use the technique. In this way the learner's attention is diverted away from the exercise of the technique so that the technique becomes integrated into their thinking.

4

MATHEMATICAL ACTIVITY

The point of setting tasks for learners is to get them actively making sense of phenomena and exercising their powers and their emerging skills. However, it is important to look beyond the mere setting and execution of tasks. So this chapter considers ways in which learners' awareness of their mathematical powers and of mathematical themes can be used to engage them in activity that leads to learning.

We distinguished between task and activity in Chapter 1: teachers set tasks which generate activity by learners. The activity itself is not learning: it is just doing tasks. But it is in the course of this activity that learners encounter mathematical ideas and themes, develop and practise techniques, and use their mathematical powers. The activity provides the basis for learning to take place.

During activity, learners engage in actions that may awaken awarenesses and sensitise them to see some aspect of mathematics differently. The actions may stimulate learners to take the initiative and explore in order to make better sense of the mathematics. The actions may also foster a desire to become competent and, in particular, to gain facility in the use of some technique. A sensitive teacher can focus learners' attention so that the activity contributes to learning.

The intentions behind tasks are various:

- they may be intended to initiate entry into some mathematical topic;
- they may be designed to provide a context in which new ideas can be encountered or ideas that have been met previously can be practised.
- they may be intended to act as revision and consolidation, or to prompt reflection and integration through an overview.

Each of these objectives emphasises one of the three aspects of mathematical activity (behaviour, emotion, awareness) discussed in Chapter 2, while drawing on the others.

4.1   ACTIVITY AND ACTION

As you saw in the introduction to this book, the distinction between task and activity suggested by Christiansen and Walther (1986) makes it possible to think about the design of tasks that have the purpose of generating activity, and not to get confused between teacher intentions and what learners actually do.

Teachers have always engaged learners in activity. Sometimes the activity is akin to that of an apprentice doing preparatory work in the presence of a master (Brown, Collins and Duguid, 1989); sometimes the activity involves learners doing repetitive exercises that follow the format of a worked example; sometimes the activity involves memorising rules and then applying them to exercises or practical situations; sometimes the teacher guides an exploration that enables learners to experience mathematical discovery and creativity.

But what is mathematical activity? It must be more than learners busily getting on with something or producing pages of written work. Although learners may be happily engaged in social interaction through discussion, or fully occupied in using scissors or drawing up tables, they may still not be undertaking mathematical activity. On the other hand, they may be sitting quietly, apparently staring out of the window, and yet be thinking deeply: this could be mathematical activity. For learners' activity to be mathematical, there have to be elements of mathematical thinking, and there also has to be some development of the learners' experience and awareness. It is not sufficient that tasks are completed more or less correctly.

Soviet educational psychologists such as Leont'ev, Galperin and Davydov (see Wertsch, 1981), who developed Vygotsky's ideas about teaching and learning, suggested that the *actions* carried out by learners need to be distinguished from learners' *activity*. For them, an *action* was something done with a precise goal in mind, whereas activities had motives that might be less precise; that is, actions relate to the micro level, and activities to the macro level. Some examples of actions and activities are set out in Table 4.1.

Table 4.1

Action	Activity
Using specific coordinates to locate or display objects.	Plotting points to see if they lie on a straight line, or finding a grid reference.
Organising information in a table or list.	Determining whether a dice is biased.

As was suggested in Chapter 3, activity only arises when there is purpose and intention. Leont'ev, as part of the general activity theory developed from Vygotsky (1978), came to the conclusion that activity is *action with a motive*, and so is 'a process which is always initiated by and interpreted in the perspective of a motive' (Christiansen and Walther, 1986, p. 255).

The most important differences between various activities are due to the differences in the motives of the people carrying them out. People form images in their minds about things they want to do, and these images, in turn, help them to see themselves in relation to the world: in this way they acquire motives. At the macro level, activity results when a person acts to fulfil a motive. At the micro level, a motive gives rise to a series of goals, which the person attempts to reach by performing actions.

To learn something by carrying out an action, it is helpful if the learner's motive is to get a sense of some underlying structure or relationship. The learner is then treating the particular object as representative of a more general class of objects.

Actions are carried out using 'cultural tools'. Cultural tools may be physical objects, such as pens, pencils, paper or rulers, or they may be symbols, such as percentage or currency signs, or they may even be formats, such as spreadsheets, for displaying information. The principal cultural tool is language, which people acquire from those around them. Such cultural tools act as mediators for learners to encounter ideas. So when children learn to use an object such as a piece of string or a ruler in order to compare two lengths that cannot be juxtaposed, they are using a cultural tool as an intermediary in the actions they are carrying out.

On the basis of a rather different source, which can be traced back to ancient India (Bennett, 1966; Raymond, 1972), action can be viewed as requiring the presence of three elements: something that initiates and *acts* upon something else, which in turn *receives*, accepts or even attracts being acted upon, all made possible by the *mediation* of a third element, impulse or force. The result of an action is a product (see Figure 4.1). In other words, when learners manipulate an object (physical, mental or symbolic), they act upon that object. These actions are mediated by a cultural tool and produce a product.

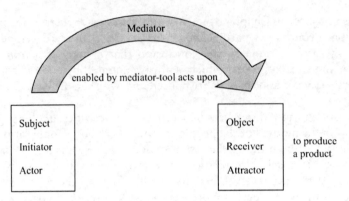

Figure 4.1   Three elements required for action.

Four examples of learners' activities and actions are given below to illustrate these ideas.

Example 1

The activity is using a ruler to compare two lengths. The motive for the actions is to compare lengths.

The learner acts upon the first length, mediated by the ruler (a cultural tool), to produce a measure, represented by a number. The learner now carries out a second action by applying the ruler to the second length and again produces a result in the form of a number. The learner then begins a new action by acting upon the pair of measurements, using the culturally mediated tool of 'base 10 arithmetic' to compare the lengths and specify 'longer', 'shorter' or 'the same'.

Example 2

The activity is extending a sequence of numbers. The motive for the actions is to extend this sequence.

The learner acts to produce a table in which successive differences are recorded. In this case the table is the cultural tool that mediates the action.

Example 3

The activity is carrying out a task involving interpretation of data. The motive is to make sense of the data, and this guides the learner's actions.

The learner distinguishes and selects relevant numbers, does percentage calculations, and then reinterprets the answers in the given context. The learner uses the cultural tool of percentages on the numbers (which are cultural tools themselves) in order to make sense of the situation.

Example 4

The activity is using a computer program. The motive is to find treasure hidden on a coordinate grid.

The learner's actions are to enter pairs of coordinates on the computer. The cultural tools include the computer and the coordinate system. After each attempt a clue is given, which shows the distance of the learner's current choice from the treasure.

The learner's actual motive is to find the treasure and so complete the task. However, when the teacher questions the learner it transpires that that they have been ignoring the clues and that their coordinates are guesses, more or less at random. The teacher's intention was that the learner would, by reflecting on their choices and the clues, devise a strategy for finding the treasure. The learner's actions are the correct ones, but their activity is not the one envisaged by the teacher.

An important reason for distinguishing actions from activity is illustrated by Example 4. Merely because the appropriate actions are being carried out (in this case, pairs of coordinates entered into the computer), it does not follow that the activity is a valuable one: the same actions could be used in many different activities. It is through participation in actions that learning takes place. But if learners misunderstand the intentions behind the actions, or if they look for shortcuts, or even copy from others, they will miss the effect of the actions on their awareness.

It is evident that there is nothing inherent in a task or its associated activity that will result in the intended learning. What is crucial is the learner's perception of the task and of the activity. Activity becomes productive when learners use their initiative to engage with mathematics. For a task to be appropriately challenging and afford possibilities for learners to encounter something significant, there must be a gap between the learners' current state of knowledge and the goals that they adopt. There must also be suitable resources to hand, which are capable of mediating movement towards those goals.

Activity theory suggests that there are three ways in which learners perceive and control their own activity:

> First, by their '... internal maps of the world—their structure of already acquired knowledge, concepts, self-understanding, systems of values and norms, positions, viewpoints ...';
> second, by their internal representation of what it is they wish to do;
> and third, by using feedback '... to compare data collected in the environment with answers established by means of [their] internal maps ...'
>
> Christiansen and Walther, 1986, p. 257.

Each of these will cause modifications in the learner's view of the task and probably in the task itself.

Christiansen and Walther also suggested that the key aspects that affect the learner's activity in class are:

> ... the individual's interest in the task, his motivation for acting, his attitudes towards the teacher and the school, his conceptions of learning and of mathematics ... and ... whether he *reflects* on his actions and on his own learning.
>
> Christiansen and Walther, 1986, p. 262

In Section 4.4 below these motivational aspects are developed more fully, but first we examine what constitutes mathematical activity.

4.2   TYPES OF MATHEMATICAL ACTIVITY

What does mathematically desirable activity consist of? In the main, it involves learners using their natural powers to solve problems or explore situations. Every learner possesses natural powers whose use and development are vital for learning mathematics. These powers were mentioned in Chapter 3; here they are described more fully.

NATURAL POWERS

The kinds of powers that are relevant for learning mathematics are ones that learners will have demonstrated by the time they arrive at school. Learners have the innate ability to emphasise or stress some features and to ignore others, enabling them to discern similarity and difference in many subtle ways. They can also specialise by recognising particular instances of generalisations, and they can generalise from a few specific cases. In addition, they can imagine things and express what they imagine in words, actions or pictures, together with labels or symbols.

Six pairs of natural powers are now discussed. The pairs are: stressing and ignoring; specialising and generalising; distinguishing and connecting; imagining and expressing; conjecturing and convincing; organising and characterising. In each pair, the two powers are 'opposite sides of the same coin'.

Stressing and Ignoring

All human senses work by stressing some features and, consequently, ignoring others. Thus a person might pay attention to (that is, stress) the commentary on the television but ignore someone who is talking in the room, or stress people's body posture but ignore what they are saying.

As soon as stress is placed on something, it raises questions about that feature. In mathematics, this frequently leads to asking whether the feature could be changed in some way. For example, consider the statement 'The product of two consecutive numbers is divisible by two'. Stressing the word 'product' here might prompt the question 'What about the *sum* of two consecutive numbers? ' However, stressing 'two numbers' might lead to asking 'What about *three or more* numbers?' and so on. Observations of this kind led Gattegno (1970, 1987) to identify stressing and ignoring as the basis for generalisation and abstraction.

Specialising and Generalising

Generalising is a natural consequence of stressing and ignoring. But it is also much more than that. It can be part of everyday life: people seek the general through the particular, and conversely they see the particular in generalities. This wider use of generalising shows itself in remarks such as 'He is a reckless driver' or 'Magpies always seem to occur in pairs'. Alfred North Whitehead, mathematician and philosopher, expressed this as:

> The progress of science consists in observing ... inter-connexions and in showing with a patient ingenuity that the events of this ever-shifting world are but examples of a few general connexions or relations called laws. To see what is general in what is particular and what is permanent in what is transitory is the aim of scientific thought.
>
> Whitehead, 1919, p. 4.

Of course, in everyday life most generalisations have exceptions, as you quickly find when you make a general statement and someone contradicts it with a specific case. This indicates that it is important to test generalities by trying out a particular case—in other words, by specialising. In mathematics, by trying one or more special cases, you can get a sense of what the generality is about, and maybe move towards a new generality. As an example, consider the general statement 'Take any three whole numbers: there will be two of them whose sum is even'. To make sense of this, it helps to check the validity of the statement by choosing three particular numbers, and then, to get an idea about why the statement is always true, try out other cases.

Distinguishing and Connecting

Operating in the world requires people to distinguish 'things' from the background so that they can become aware of properties and relationships between those properties, and so on. For instance, in order to count, it is necessary to be able to distinguish what is to be counted (you cannot count raindrops in a puddle). Similarly, multiples of, say, 3 can be distinguished from all of the counting numbers; it is then possible to go on to distinguish those multiples of 3 that are even and to relate them to those that are odd, noting that they occur alternately (and connecting them to multiples of 6). Once things have been distinguished, it is equally important to recognise what is similar and what is distinctive, again a result of stressing and ignoring.

Imagining and Expressing

The power to imagine is vital to the way that human beings function, and this carries over into mathematics. Patterns, whether of numbers, shapes or algebraic expressions, are imagined as continuing even though only a few examples are actually given, and it is this imagined continuation that gives the pattern its significance.

However, an imagined pattern is solely inside one's head until it is expressed, so being able to articulate a pattern, whether in words, diagrams or symbols, is an essential mathematical technique.

Conjecturing and Convincing

Much of what people say is, at best, conjecture. Often someone says something with the intention of seeing how it will go down, and with the expectation that it can be looked at critically, tested against experience and modified if necessary until other people are convinced.

For example, when two line segments are imagined, a first conjecture might be that the line segments either intersect or do not. But that assertion is open to question: what if the segments do not actually cross, but one has an endpoint on the other? And what if that endpoint is at an endpoint of the other? And what if they overlap? The conjecture might then be expanded to give several conjectures, or perhaps the meaning of 'intersect' might be examined more closely. In mathematics, it is important to be aware of the status of assertions: have they been established as always true, sometimes true, never true or are they still just conjectures?

Organising and Characterising

It is natural for people to organise things (think of your kitchen) in order to reduce the effort required to remember. It is also natural to characterise objects by seeing their features ('Oh, that's a labrador'). In mathematics, characterising objects can be a way of saying that they are all 'of the same kind'. In some senses, all even numbers behave in the same way, as do those numbers that give a remainder of 1 when divided by, say, 3. You can also characterise all problems of a given type that yield the same answer.

HOW TASKS PROMOTE ACTIVITY

Tasks can be effective in producing useful activity in various ways:

1   A task can ask learners to *imagine* something, perhaps supported by a diagram or computer animation, but with extra constructions imposed mentally. The aim is to evoke in learners some aspects of the images and connections that are usefully associated with the topic in hand.

> For example, 'imagine a triangle and make it as extreme as you can'; or 'imagine a number between 2 and 3, with at least three decimal places including the digit 7, to place it on the number line'.

2   A task can ask learners to *generalise*, perhaps by explicitly exposing them to a sequence of particulars from which they can express a generality, or by getting them to consider dimensions-of-possible-variation, or by encouraging them to see through the particular to a generality.

> For example, when given the problem 'A rectangle has sides of length 7 and 3. What is its area?', the task can continue 'What dimensions-of-possible-variation can you think of, and what is the corresponding range-of-permissible-change?'

3   A task can *surprise* learners by displaying a situation in which something unexpected happens, and then requiring the learners to provide an explanation.

> For example, a ribbon is tied round a box by placing it over one corner, then under the adjacent corners, passing beneath the box, and then back up over the corner opposite the first. 'In what position is the ribbon the shortest?'

4   A task can *create possibilities* for learners to be creative, to experience generality, or to encounter freedom-and-constraint. This can be done by changing a 'doing' problem into an 'undoing' one, or by asking 'If

this is the answer to this type of problem, what was the question?', or by reversing the data that are given and the data to be sought.

> For example, 'If 24 is the answer, what could the question have been ... in the context of greatest common divisors, ... in the context of least common multiples, ... in the context of factorising, ... in the context of areas made up of rectangles?'

5   A task can get learners to *produce examples*, starting with a wide degree of freedom, and then, by imposing a sequence of constraints, getting the learners to consider the range of possibilities and how much freedom there is at each stage.

> For example, think of three numbers; now think of three numbers whose sum is 18; now three numbers whose sum is 18 and whose product is as large as possible.

When learners are given a task, there are important phases of activity which come and go. There may be periods of excitement as learners engage with some problem. There may also be periods of frustration, when conjectures do not work out or when ideas are few and far between. There will be times for sitting quietly and thinking, or for listening to others discussing their ideas.

Getting started is rather like the flaring up of a match when first struck. But to be sustained and useful, the match needs to ignite something more substantial. This corresponds to switching initiative from the teacher to the learner. Something equivalent to a bellows-action may be needed before the emotional energy is sufficient to get learners fired up.

The image of fire can be developed further. As a fire dies down one can put on more logs or coal; this is akin to the teacher extending or varying the task-situation in order to sustain mathematical activity once learners understand the real question behind the initial task.

As a result of undertaking tasks, learners may appreciate more deeply some aspects of mathematical thinking in general, and the topic in particular. The aim is for awareness to be educated, in the form of powers developed, themes extended and exemplified, thinking processes exercised, concepts encountered and articulated, technical language embraced and employed, and so on.

4.3   THE FLOW OF ACTIVITY

The core of teaching mathematics lies in the interactions between teacher, learners and mathematics within the classroom and institutional milieus. It is through these interactions that learners' expectations are reinforced, their self-image as mathematicians is established and their motivation to make sense is either affirmed or dissipated.

Learners' activity will not be uniform throughout a single lesson or series of lessons: there will be a flow from one kind of activity to another, guided by interactions. It is important that these interactions are not random, but enable some development in the learners' activity.

We now examine three frameworks for thinking about such development. They are:

- Manipulating–Getting-a-sense-of–Articulating;
- Doing–Talking–Recording;
- Scaffolding-and-Fading.

You met the first of these in Section 3.2 above in relation to tasks.

MANIPULATING–GETTING-A-SENSE-OF–ARTICULATING

Activity by learners can be analysed in terms of how they *manipulate* objects and thereby *get-a-sense-of* structures and/or relationships, which eventually they are able to *articulate*.

As an illustration of this framework, consider the kind of activity in which learners can be prompted to use apparatus in such a way that an underlying structure is revealed. For instance, when exploring even and odd numbers, learners can be asked to 'build' some numbers by *manipulating* Multilink cubes. Each even number will be built with two identical rows, and each odd number will have an extra cube in one of the rows. By fitting together different blocks of cubes, learners can *get-a-sense-of* the structure of adding even and odd numbers. By talking to themselves and to others about that structure, learners can *articulate* general statements of the type 'Adding two odd numbers gives an even number'.

There are several points to note about this example. First, the apparatus is such that learners can manipulate it with confidence. Second, that the constructions have a purpose to see what is going on when adding even and odd numbers. Next, because each learner makes their own particular Multilink construction, there will be a large number of different constructions in the class, so learners can see that each is an example of the more general set of even or odd numbers. Lastly, it is easy for learners to construct other examples to test the conjectures being made.

Once learners have established general statements, such results can be used as 'objects' in future manipulations: thus learners use the fact that 'adding two evens gives another even' in other problems.

What emerges is a spiral of development, consisting of the following stages:

- the manipulation of physical, mental and symbolic objects in order to get-a-sense-of structures, relationships and/or generalities;
- an on-going struggle to bring these to articulation, even though the sense-of changes as articulation develops;
- a shift in the way that learners see things, which enables them to manipulate with increasing confidence what previously seemed abstract to them.

These steps will be repeated by the learners, and so result in an on-going spiral of development.

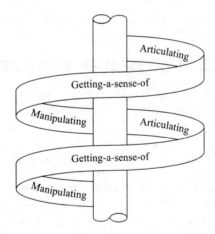

Figure 4.2   Spiral of development.

When learners find themselves uncertain about what something means, they can decide for themselves to backtrack to a point where they feel more confident, and use examples they understand in order to get-a-sense-of what is going on.

The Manipulating–Getting-a-sense-of–Articulating framework was first produced as practical advice for teachers. It can serve as a reminder to teachers that the purpose behind learners using apparatus or doing examples is for them to get-a-sense-of structure and underlying similarities, and then for learners to articulate these as conjectures and express them by means of diagrams and symbols.

The ideas behind the Manipulation–Getting-a-sense-of–Articulating framework structure have close similarities with Dienes' insistence that understanding grows from manipulating things (Dienes, 1963), and with the three modes of representation identified by Bruner, described in Chapter 1: *enactive* (manipulating physical objects), *iconic* (mental images and sense-of) and *symbolic* (using symbols).

To appreciate the relevance of Bruner's modes, consider the learning of place value. Here an *enactive* representation would involve using, or manipulating, bundles of ten sticks. An *iconic* representation would involve drawing pictures of bundles of ten when beginning to record transactions concerned with place value. A *symbolic* representation would involve using one 'thing' to represent another where there is no clear connection between the two: with place value, this would require standard symbolic representations of numbers greater than 10.

Of course, the boundaries between Bruner's modes of representation are fuzzy. Numerals are symbols, but they become so familiar that they turn into objects that can be manipulated when doing arithmetic. The 'manipulating' in Manipulating–Getting-a-sense-of–Articulating is intended to have an even wider sense, in that learners are involved in manipulation when they specialise some generality to instances which they can handle with confidence, and then use those instances to re-formulate the generality for themselves. In this sense, Specialising and Generalising, together with Conjecturing and Convincing, as well as Bruner's Enactive–Iconic–Symbolic modes, and the Manipulating–Getting-a-sense-of–Articulating spiral all address the same thing in different language and from slightly different perspectives.

DOING–TALKING–RECORDING

As you have seen, Manipulating–Getting-a-sense-of–Articulating constitutes a framework for relating learners' activity to their development. Running in parallel with this activity is what learners are actually doing—their actions. There are frameworks for summarising these actions. One such framework for describing learners' overt behaviour is referred to as *Doing–Talking–Recording*. This framework was formulated as a reminder that when teachers rush learners from 'doing things' to producing written records, the teachers are likely to miss out the vital ingredient of getting learners to talk about what they have done.

When learners are *doing* calculations or manipulating objects so that they can get-a-sense-of some idea or process, the teacher can facilitate this by structuring tasks using the Doing–Talking–Recording framework. Learners are induced to talk about what they are doing and sensing (whether seeing, imagining or considering) before making a written record.

Talking in mathematics classrooms has been much researched and theorised (Sfard *et al.*, 1998). However, its importance and role are most evident not as an end in itself, but as a component in the Manipulating–Getting-a-sense-of–Articulating spiral of development, making use of learners' natural powers of sense-making and mathematical thinking. It is not enough for the teacher merely to encourage discussion. To be useful, talk needs to be a strategy directed towards a more focused mathematical aim. So, for example, conjecturing arises not just from doing, but from doing, talking and recording; the desire to make sense emerges as a desire to explain when there is disagreement or surprise.

Teachers can be tempted to move learners prematurely onto *recording* their thinking so that there is some physical outcome of the work. To be effective, especially with reluctant writers, recording needs to be purposeful and to contribute to learning. This is helped by the way in which doing, talking and recording all feed into each other. Recording can arise naturally, even spontaneously, as learners become articulate about what they are doing. Their 'doing' becomes more efficient as a result of trying to record it by expressing in words, pictures or even in movement. Consequently, their articulation informs and is clarified by the process of recording what they are doing.

Additional evidence for the importance of each component of the Doing–Talking–Recording framework in relation to Manipulating–Getting-a-sense-of–Articulating is found in the work of Vergnaud (1983). He noticed that learners often appear to have what he called *theorems-in-action*. By this he meant that learners act as if they know a result or fact, without being able to articulate it, or even realising its existence. Thus, children spontaneously add two numbers in either order, acting as if they know that addition is commutative. The practical implications of this are that a teacher can gradually draw learners' attention to the underlying principle, theorem or fact, so assisting the learners in 'discovering' or '(re)-constructing' for themselves what they already know implicitly. Schooling can be seen as progressive formalising of theorems-in-action and of awarenesses, as well as the developing of new theorems and awarenesses.

How do learners come to have these theorems-in-action? Several theories provide possible explanations. Vygotsky's social perspective suggests that learners may pick up a social practice from more expert adults. Gattegno's view of awarenesses suggests that learners may use their own subtle and unconscious awareness of how things work in the material world. Piaget's notion of genetic epistemology suggests that learners may act out subconscious conjectures which they are constructing for themselves, supported by the school and outside environments. Perhaps no one of these perspectives has an exclusive claim: there may be elements of each involved.

SCAFFOLDING-AND-FADING

Earlier it was suggested that learners could be led into becoming more independent if teachers gradually make their prompts less and less direct. This suggestion is a particular case of a more general principle arising from Bruner's analysis of Soviet activity theory.

Vygotsky proposed that there is usually a difference between what learners can do unaided in the way of problem-solving, and what they can do when supported by someone more experienced or by other learners. He called this difference the *zone of proximal development* (ZPD) and described it as

> the distance between the actual developmental level as determined by independent problem-solving and the level of potential to development as determined through problem-solving through adult guidance or in collaboration with more capable peers.
>
> Vygotsky, 1978, p. 86.

This notion has been interpreted far beyond Vygotsky's original intentions, to the point where some authors talk about 'working in the ZPD' as if it were some place or space.

Bruner (1986) and colleagues, importing Vygotsky's ideas, introduced the term *scaffolding* to describe the role played by the teacher when supporting a learner in doing something they could not yet do for themselves. The teacher might play a supportive role which would not only enable learners to solve problems they could not do on their own, but to learn from that experience. The image associated with scaffolding is meant to be one of temporary support provided by the teacher, which is later withdrawn.

Unfortunately, scaffolding is another idea that became so popular that its use expanded to such an extent that it became virtually meaningless. Despite this, when researchers look specifically for something they can identify as scaffolding, it proves remarkably elusive (Askew *et al.*, 1997).

The important awareness that underlies scaffolding is best seen by combining the concept with the notion of fading, to give the *Scaffolding-and-Fading* framework. The idea is that a teacher might ask a certain type of question such as 'Can you give me an example?', in several different contexts. The effect of such a question is supposed to be that it assists those learners who did not think to ask themselves that question. Thus the teacher scaffolds the learner's work by being 'consciousness for two' (Bruner, 1986, p. 75). At some point the learners might be expected to become used to the question and know what sort of an answer is wanted. The teacher's contribution can then start to fade, becoming more indirect and involving meta-questions such as 'What did I ask you yesterday in a similar situation?' or 'What did you do last lesson in a similar situation?' Such prompts can become less and less direct, until eventually learners start to use the questions on their own. In order to indicate what sort of behaviour is being sought, the teacher might want to draw attention to instances where learners might ask such questions of themselves. In this way, not only do learners become less dependent, but they also learn how to learn, and consequently a constructive way of working emerges.

As learners take over and internalise one form of prompt, the teacher can direct their attention to other matters, and perhaps engage in a similar Scaffolding-and-Fading process with another prompt. One difficulty for teachers is knowing when to stop scaffolding and fading. If their use of a prompt does not fade away, then they are training learners to be dependent and not to learn effectively.

4.4   ACTIVITY AND EMOTION

Vygotsky suggested that the best task for initiating suitable learner activity is one which a learner is eager to engage with.

How is this engagement to be achieved? 'Eager to engage with' implies an emotional thread; indeed, learners are likely to engage in appropriate activity when they are trying to make sense of some situation that intrigues them. You saw in Chapter 3 that tasks which surprise or intrigue learners are likely to harness emotions more readily than tasks which only call upon them to select the correct technique and apply it accurately. There are many types of task that are suitable to engage learners: they can be attracted into activity by tasks that involve playing a game, or acting something out, or manipulating some object on a screen, or watching an animation, or discussing a poster. When learners are intrigued, they are more likely to sustain their activity by trying to identify an underlying structure, or to generalise it to other contexts.

Of course, not everyone will be intrigued by the same thing. But most learners can be drawn into a task that presents them with some sort of *disturbance*—whether a surprise, a contradiction of expectation, or even an annoyance with something. They are then more likely to want to engage in sustained activity in an attempt to resolve the problem.

Many writers have pointed to the benefits of learners encountering some sort of disturbance to spark their interest. Some have advocated methods for doing this, and it is to these we now turn.

HARNESSING EMOTION THROUGH DISTURBANCE

We look at two ways in which learners' activity can be promoted by producing some kind of disturbance. These both involve creating 'cognitive dissonance' in learners, that is, presenting them with instances that conflict with an already established position.

Torpedoing

Davis (webref) coined the term 'torpedoing' to refer to a teaching strategy that deliberately leads learners to make a conjecture which then has to be modified in the light of further evidence. For example, he devised a task for 8- to 12-year-olds in which they had to find numbers to put into boxes so as to make statements, such as the following, true:

$$(\Box \times 1) + 11 = 77 \quad (\Box \times 1) + 23 = 39$$

The learners were then asked to make up a comparable example of their own.

Once the learners had gained facility at doing those, they were asked to do these:

$$(\Box \times 2) + 3 = 5 \quad (\Box \times 2) + 11 = 77 \quad (\Box \times 2) + 23 = 39$$

followed by as many more rows as necessary until everyone had generalised the pattern. If needed, Davis offered sporadic examples such as

$$(\Box \times 15) + 31 = 106$$

As ever in effective mathematics lessons, the point was not to get the answers, but to encourage learners to pay attention to how they got them. So at each stage, learners engaged in discussion of *how* they were obtaining the answers. To learn more from the experience, they also discussed what was the same and what was different about apparently similar questions.

Here is another example that involves several deliberate 'torpedoings'.

Task 1
Decide which of these numbers are even and which are odd:

3, 16, 28 844 629, 3 × 3 × 5 × 14

Make up more cases of your own, and formulate a general method for telling whether large numbers are even or odd.

Task 2
Decide which are even and which are odd:

⁻2, ⁻3, ⁻233.

Make up your own cases, and formulate a general method for telling whether negative numbers are even or odd.
Is zero even or odd? Which rule-definition are you using? Does it matter?

Task 3
Make a list of numbers that, when divided by 5, leave a remainder of 3. Put the list in order with the smallest number first. Describe all of these numbers.

Task 4
Remainders of negative numbers are always treated as positive, so, for example, the remainder on dividing ⁻3 by 5 is 2, and the remainder on dividing ⁻6 by 5 is 4. Produce a list of negative numbers that, when divided by 5, leave a remainder of 3. Describe all of these numbers.

In Task 2, learners first come up against a surprise if they have not previously thought about even and odd in the context of negative numbers and of zero. Because of symmetry, the even/odd distinction is easy to describe for negative numbers. However, there is a further surprise in Task 4 when division is switched to dividing by 5, as the even/odd symmetry is disrupted.

An extension to Tasks 3 and 4 could involve looking at other remainders when dividing by 5, and then changing the divisor 5 to other numbers.

In a similar vein, Hadas, Hershkowitz and Schwarz (2002) tried a geometrical 'torpedo' by asking a group of learners the following question:

In the figure (Figure 4.3), the base of the triangle has been trisected. Could the three angles ever be equal?

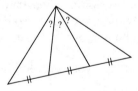

Figure 4.3   A trisected triangle.

The learners were first expected to use geometrical reasoning to solve the problem. Most learners conjectured that the angles would always be equal.

They were then asked to create such triangles using dynamic geometry software, and they quickly realised that, at best, the three angles were equal only in some positions. A subsequent search for cases of equality led some learners to state, with reasons, that it is never possible for the three angles to be equal.

The method of deliberately setting up expectations and then 'torpedoing' them can be a very fruitful way of getting learners to engage in mathematical activity.

Conflict-discussion tasks

Bell and colleagues (Bell, 1986; Bell and Purdy, 1986; Bell, Brekke and Swann, 1987a, b) noticed that learners often mix up notation for division (does 24 ÷ 3 mean '24 divided by 3' or '24 into 3'?) and that they often decide which is the appropriate meaning by a sense that 'multiplication makes bigger and division makes smaller'.

The researchers designed tasks that would reveal to learners their inconsistencies and confusions, and would lead them to reorganise their thinking. The tasks used a set of cards on which there were various symbolic expressions $(24 \div 3, 3/24, 3\overline{)24})$ and verbal versions ('twenty-four divided by three', 'three divided by twenty-four', 'twenty-four shared among three', and so on), as well as simple problem contexts whose solution required these operations. The task was essentially to sort the cards into groups so that in each group the cards all meant the same or led to the same answer. Learners produced conflicting interpretations, and often heated but useful discussion ensued. The principle behind such a task is that 'if there is an underlying misconception, then it's "better out than in"; it needs to be seen and subjected to critical peer group discussion' (Bell, 1986, p. 27).

Such conflict-discussion tasks can be used with many topics. Getting learners to sort cards or other objects into groups that are 'most alike' or which have the same answer provides a way of stimulating their engagement with mathematical ideas. Learners can be further involved if the teacher gets them to identify where they ran into uncertainty and conflict, and then to make cards that they think might produce more conflict and discussion.

4.5 SUSTAINING ACTIVITY

Energy ebbs and flows. A fresh insight can stimulate learners to test out a new conjecture, but, equally, lack of progress can demotivate them. Sustaining learners in productive mathematical activity is central to the teacher's task. It is not a matter of maintaining excitement and involvement at a fever pitch, as this is not possible—or desirable. Nor is it a matter of designing little tasks that learners will enjoy for enjoyment's sake. If learners come to depend on the teacher for excitement and drive, they will not learn to use their own powers to think mathematically, and so will not become independent and confident learners.

Once teachers have initiated learner activity, they need to be alert and receptive to learners' actions. This is the crux of teaching, because it is through the interactions of:

- learners with mathematics and teacher;
- learners with mathematics and each other;
- learners with mathematics and themselves;

that learning is enabled.

While it is sensible for a teacher to mistrust sudden exclamations by learners of 'Oh, I see' or 'Oh, is that all it is!' as these sentiments are likely to be transitory, it is in the learners' actions, as well as in their shifts in attention and in the extensions of their awareness, that the possibilities for learning are laid down. Remember that while teaching takes place *in* time, learning takes place *over* time. What learners attend to and how they view it can be changed temporarily by a teacher, but for learning to occur, a build-up of experiences, layer by layer, is needed. Moreover, learners need to integrate their experiences.

The process whereby experience and ideas are integrated in a person has been much studied by psychologists and educators. Piaget (1971) used his knowledge of biology to suggest a biological metaphor for this process. He identified *assimilation* and *accommodation* as two fundamental actions: assimilation means taking on board what you experience, while accommodation means adjusting your previous thinking to take account of new experiences. Gattegno (1970, 1987) spoke in terms of *integration through subordination*, in which techniques become automated by reducing the amount of attention needed to carry them out. In effect, the more the teacher is specific about what learners are expected to do, the easier it is for the learners to display the required behaviour without actually understanding what they are doing.

Another feature that needs to be counteracted by the teacher is the didactic tension mentioned in Chapter 3. The whole point of a lesson can be lost if the learners work out how to do the task mechanically, without having to transform or act upon their current understanding—that is, without working to assimilate or accommodate new concepts, techniques and ways of thinking.

SUSTAINING INTEREST

What are learners thinking about as they enter a classroom, or when there is a switch of topic? Learners of every age have a mixture of desires and concerns that affect their motives and hence their activity in class. These can be as general as concerns about life outside school and about themselves as individuals, or desires to be challenged and occupied. Do they eagerly anticipate finding something out? Do they become defensive, looking for avoidance strategies in order to delay boredom? Do they seek to lose themselves in plenary sessions by copying others and hiding what they do not know? Do they respond to being called upon to imagine and to express, to specialise and to generalise, to conjecture and to convince?

On starting a mathematics lesson, what sense do the learners have of what they will be doing? Is there some sustained activity for them to continue, or do they always expect a fresh start to each lesson? The practice of starting each lesson on a new task or topic in order to 'keep learners interested' may lead to a descending spiral as learners claim not to remember the last appearance of a topic, so the teacher reminds them and the learners recall even less. They may stop using their powers of mental imagery to re-enter what they were doing and to imagine what they will be doing in the next lesson, and may foster a sense of mathematics as an endless smorgasbord of disparate topics and techniques that have to be memorised. However, if learners are expected to remember where they left off and to pick up their work from that point, then they are likely to respond by meeting such an expectation.

Learners' self-esteem is often quite fragile, but it is not strengthened or enhanced by the teacher trying to make everything simple for them. They know perfectly well when they are being given over-simplified tasks whose completion does not warrant special praise. Learners can gain pleasure from tackling and overcoming challenges but only when these are seen to be both significant and tractable. As has been remarked elsewhere, learners' emerging awareness of their natural powers to think mathematically, as well as the use and development of those powers, are motivating in themselves.

Claxton has drawn attention to characteristics of the temperament that supports learning most effectively: resilience, persistence, a playful disposition and a willingness to learn with others. As he said,

> ... you can have the best toolkit in the world but if you're really frightened you won't dare to use your tools so the first aspect of temperament is what people are referring to as resilience. That's stickability. The ability to hang in there with something when it gets difficult.
> Claxton, webref.

SUSTAINING MATHEMATICS

Another way in which learners can sidestep productive mathematical activity is when they want to finish the task they have been given with the minimum of investment. When learners finish a rich task quickly, they do not really make contact with the mathematical thinking processes involved nor with the mathematical themes; also, they do not gain facility in using the techniques that the task calls for.

One of the ways in which learners can be helped to avoid their activity becoming routine is to get them to see an individual task as representative of a whole class of similar tasks. This requires a shift from 'doing a question' to seeing the question as representative of a class of questions. As a first step to this end, learners can use the idea of dimensions-of-possible-variation to provide a way of talking about features of a problem that they can change. For instance, the teacher could ask the learners to:

- change any of the explicit numbers that act as data;
- change implicit or structural numbers;
- change which data are given and which are to be found;
- change the context of the data;
- change the order of presentation of the question.

Subsequently, learners can be encouraged to ask themselves questions such as:

- What would similar questions look like? Can I make up my own questions?
- What dimensions/features/aspects could I alter and still be able to do the task?
- What is the same about all the examples, and what is different about them?
- How would I recognise a task like this in the future?
- What is the essence of this question? What features could be changed without changing the question?

It is one thing to solve a particular problem once; another thing to be able to solve any problem of a specified type when encountered; and quite another thing again to solve all such possible problems and produce a 'formula'. The further learners get in this development, the more likely they are to remember a technique and to use it later when appropriate, and the more secure they will be when they sit an examination.

Some learners will work quickly and may need to be challenged; others will work less quickly and may need the teacher to support them by listening and encouraging them to use their powers. To ensure that the faster learners engage in appropriate mathematical activity, they can be taught how to extend tasks and make them more complex through generalisation and through exploration of dimensions-of-possible-variation. Almost any task can be extended, often in challenging ways. Some useful strategies for extending tasks are the following:

- Where there is a choice of answers (as in 'find a number between 2.352 and 2.371'), try to count or to classify all the possible answers (in this case, how many numbers there are between 2.352 and 2.371 with three decimal digits).
- Where there is a single answer (as in 'find the average of 2.352 and 2.371'), reverse the task by converting a 'doing' into an 'undoing' problem (in this case, find other pairs with the same average or find three numbers with the same average).
- Generalise questions (as in 'describe how to find a number between any two given decimal numbers of any specified number of decimal digits').
- Where there are several questions of the same type, construct easy, hard and general questions of that type, and consider what other questions of that type will give the same answer, and also what other answers are possible to such a question.

The strategy of converting a 'doing' into an 'undoing' problem can produce quite difficult problems. Sometimes there are several questions, even a whole class of possible ones, all of which give the same answer. As a step in this strategy, learners can be asked what features the question has to have so that the answer has some relevant property. For example:

> What features must equations of the form $2x = 3$ or $5x = {}^-15$ have if the solution is to be positive, or zero, or an integer?

This is also a way of characterising types of answer.

The principal use of algebra by mathematicians is when they generalise a problem to a class of problems, or find a general solution. Even young learners can experience this process by applying it to word problems. They can generalise from a few problems to a class of them, then formulate a method for solving any such problem and possibly even present a formula for this.

There are various devices that teachers can use to 'keep learners involved long enough' so that they get something from participating, beyond another filled page in their exercise books. Learners can be invited to:

- construct a game in which the players use strategies based on ideas from the topic;
- make up their own examples to explore relationships and, in passing, do calculations that they need to practice;
- characterise the possible answers that can arise from a given type of question.

After a period of working on a problem, learners can offer their ideas for making progress in a plenary discussion so that others can pick up on fruitful ideas before spending too much time on less fruitful approaches. This reinforces the learners' sense that they can participate by making choices and, by making choices, enjoy participating. Similarly, after a period when learners have formulated their own questions, they can pool their ideas and decide which questions look tractable and interesting to pursue.

4.6 STANDING BACK

The previous sections considered the ways in which a teacher can direct learner activity so that learners do more than just complete tasks. As has been stressed, it is possible for learners to engage in activity while missing the point of it. They can copy and complete a table of values without ever becoming aware of the relationships in the table; they can draw a diagram as an exercise in copying, without ever paying attention to how the diagram is built up, what can change and what relationships are invariant; they can do a suite of exercises, get most or all of them correct, and yet not learn anything from the effort.

What matters most in lessons is not the answers that learners have obtained, but what they have experienced. It is tempting to add 'what they have learned' but, as has already been pointed out, learning is a maturation process and takes time.

Although people say 'I will never forget that', most things *are* forgotten; something which seems unforgettable (such as recognising letters or numerals, knowing how to add fractions or to subtract three-digit numbers) can fade away all too quickly. Activity in which learners meet ideas and techniques is not sufficient for those ideas and techniques to be recalled when they are needed. Learners have to integrate this information in some way. But how is this done? Educators and researchers who have tried to explain just how people learn mathematics all favour some form of crystallisation, in which many particulars are subsumed under one generality. In Piaget's terms, learners have to *assimilate* and *accommodate*.

Although integration seems to take place mostly while people are asleep at night when their minds are making sense of all the things that have happened during the day, integration can be supported by a different kind of activity—one that involves learners in standing back from doing a task, and shifting attention from the doing to the how, why and when of doing. This requires learners to become aware of what they have experienced, and to abstract or generalise it. One of the simplest ways of fostering this process is for the teacher to ask learners questions that seek to bring their awareness of specific mathematical themes, techniques and strategies to the surface, as well as increasing their awareness of their own natural powers. Some questions of this kind are:

- What could you tell people at home about what you did today?
- What might you have done better? More efficiently? What was effective and what was not? What might you want to do differently next time?
- What did you notice about how you worked on the task?

At first, learners may not be able to remember very much about their activity. They may have been immersed in the doing and may not have really been aware of what they were doing or why. But repeated use of such questions can develop learners' ability to re-enter recent experiences. Learners can then gradually take over these questions for themselves.

REFLECTION

The questions above are examples of prompts that encourage reflection. Although 'reflection' is a much over-used word, the act of reflection, that is, of standing back from the immediate 'doing' and considering what you have been undertaking, is a major way in which experience contributes to the maturation process of learning.

When learners summarise what they think a lesson has been about, whether to themselves or to others, the process can help to consolidate their experiences. It can both reveal what learners are aware of, and indicate where they need to do more work and/or have more support.

One of the most fruitful ways in which a teacher can get learners to start summarising is to ask questions of the form 'What is the same and what is different about ...?' Coles (Coles and Brown, 1999; Brown and Coles, 2000) has shown how such questions can develop into a powerful way of working which learners can adopt and use spontaneously amongst themselves and for themselves, leading to more efficient and effective learning.

An emphasis on 'reflection' as a core contribution to learning might seem surprising. It could be argued that most of us have learned a very great deal without having reflection as an explicit component of our way of working. There are, in fact, many other components. Hilary Evens drew up the following list after listening to Open University students talking about changes in their mathematical understanding:

> read, recall, recap, recognise, recollect, reconstruct, record, redo, re-express, refer (back), reflect, reformulate, refresh, register, regurgitate, rehearse, reinforce, relate, relearn, remember, remind, reproduce, reread, result, retain, review, reverse, revise, revisit, rewrite.

The prefix *re-* in reflection implies looking back, but a number of the words in this list do not involve looking back; some represent actions that are performed when doing a task.

It is useful to distinguish between those parts of lessons in which the teacher explicitly aims to get learners reflecting on what they have done and the informal reflection that goes on naturally during a task. The latter occurs whenever a learner is trying to see how something is done, or is making comparisons between what has been achieved and what was intended, or is gaining feedback and using this to modify what is being done. Learners will carry out such actions without prompting in all kinds of circumstances, and many of them do so while working in mathematics lessons. Indeed, getting-a-sense-of a concept happens spontaneously.

This kind of informal reflection is a key to much learning and, as you have seen, it is possible for teachers to devise tasks that are likely to promote such thinking. For example, a teacher might get learners to pay attention to how they 'do the next case' or 'fill in a table entry', or how they test a guess. Examples can be used to shift learners' attention from getting answers to how those answers are achieved.

Informal reflection is much more likely in a classroom ethos where explicit reflection is also encouraged, and where learners are encouraged to make choices, use their initiative and make mathematical sense of tasks through the use of their powers.

ENDING

In any classroom, learners will be working at different speeds. Consequently, whenever work is brought to a halt, whether for a change of energy or at the end of a lesson or at the end of work on a particular topic, some learners will feel they are being interrupted.

It is useful to compare this situation with that of mathematicians when they have to leave off work on a problem. As mentioned earlier, they have a very deliberate strategy. They make notes of current conjectures, of examples that illustrate those conjectures, and of what has already been found and justified. All of this makes it much easier to resume work at a later date, as well as serving to integrate what has been achieved and to get-a-sense-of the sequence of experiences. While these will not be suitable practices for many learners, tasks that give an equivalent feeling of rounding off the work are often appropriate at the end of a lesson or topic, and can give learners the opportunity to review what they have been doing and begin to integrate it into their thinking.

Some examples of tasks that enable learners to undertake activity which gives them an emotional sense of rounding off are:

- brain-storming sessions which essentially invite learners to think about the three threads that make up the structure of a topic and then to construct their own web diagram for the topic in hand. This promotes thinking about the topic from outside rather than just from within.
- Individuals or small groups make a poster displaying the task, what they did and what they found. It is interesting how the possibility of an audience increases the care with which learners work. Once it becomes customary to display a poster to the class, extra energy and commitment can be gained by producing a poster for the whole school or by putting the results on the Internet, where the audience is potentially the entire world.
- After each piece of work, the teacher chooses one or two learners to present what they have done so far to the others. This element of performance can add purpose to the finished product. Encouraging learners to rehearse by explaining to an adult at home what they have done can also serve to engage parents and guardians in the education of their children.

At the end of learner activity, what matters is not the answers to specific questions, and perhaps not even the methods for answering different types of question. What really matters is that learners develop a sense of confidence—confidence that they can tackle problems they have not seen before, that they have ways of learning concepts as well as techniques, that they have ways of dealing with being stuck on problems, and that they can reconstruct techniques and links they have encountered in the past. Above all, it is important that learners realise that learning is an on-going process, and that the only limits are the limits they impose upon themselves.

If these sentiments are accepted, then activity concluded with ticks and crosses is a waste of energy, but activity concluded with personal and collective summaries, constructions of significant and generic examples, and articulations of general principles will lead to learning.

Learners who have begun to reconstruct and to explore how to tackle novel problems for themselves have learned something worthwhile. However, learners who have become skilled only in questions of a particular format and style may be thrown by even the smallest deviation. Boaler (1997) found this when she studied two schools in similar socio-economic settings, one with a formal curriculum and individual work, the other with exploratory group work. Learners in the latter setting did at least as well as the others, but also appreciated the potential use of mathematics and mathematical thinking outside the classroom; in addition, they seemed to be more flexible in the range of questions that they could tackle.

SUMMARY

Activity by itself is just activity. Indeed, being immersed in doing something can actually block awareness of what is being done and how it is being done, and can obscure the general principles that are being manifested or activated. Just because learners are engaged, it does not mean that they are learning mathematics; just because they are making sense of what they are doing, it does not mean that they are learning. Since learning is a process of maturation, it is useless to try to force learning to happen.

This chapter has put forward a number of useful distinctions and frameworks to help in thinking about development in learner activity:

- activity and actions;
- how tasks promote activity;
- Manipulating–Getting-a-sense-of–Articulating;
- Doing–Talking–Recording;
- Scaffolding-and-Fading, and the zone of proximal development;
- harnessing emotion through disturbance, torpedoing and conflict-discussion tasks.

Activity is initiated most effectively when there is some sort of disturbance, in the form of a surprise, a contradiction of expectation or even an annoyance. Questions for learners that reflect the types of question which mathematicians ask themselves are more likely to capture attention and provide an appropriate sense of mathematics as a disciplined form of enquiry rather than as a narrow range of stylised exercises.

Expressing themselves to others and listening to what others say with a view to testing it out on examples both contribute to sustained activity which keeps the learner immersed long enough to gain sufficient exposure to the ideas, language and techniques that make up the topic in question. Learners achieve consolidation and mastery most effectively through practice rather than through mindless rehearsal; moreover, the meaningful exercising of new skills in order to get-a-sense-of something that transcends the skills being automated is important.

Ultimately, learning depends on reflection, which may be implicit, but can be explicitly promoted through various strategies.

5

INTERACTION

In Chapter 4, one of the central problems faced by teachers was identified: how to ensure that learners' activity is motivated by a desire to engage with and learn mathematics. The quality of a learner's activity depends on many different factors in complex ways. As shown in Chapter 3, the purposes and intentions of author, teacher and, especially, learners are crucial. Learner attitudes and intentions are sensitively dependent on the atmosphere that emerges as teacher and learner begin to engage with each other. The main way in which a teacher can help learners have an appropriate motivation for their activity is by interacting with them.

The totality of teacher–learner interactions is vast and complicated. This chapter concentrates on just a few aspects of the verbal interactions that take place while learners are working on tasks, and which can profoundly influence learners' attitudes towards themselves, towards mathematics and towards learning itself.

Section 5.1 takes an overview of the issues involved in verbal interactions, and then Section 5.2 probes more deeply into the kinds of questions and prompts used by teachers, as well as how teachers deal with learner responses. In a similar way, Section 5.3 examines the interactions involved when teachers are telling and listening, and also considers the interactions in classroom discussions.

5.1   VERBAL INTERACTIONS

'You can tell people and tell people and tell people but if they're not ready to hear you, then there's no point'
                                        From the 1984 film *A Passage to India*.

Most verbal interactions involve people asking, telling or listening. In the midst of asking or telling, it is hard for speakers not to assume that the audience is experiencing the same thoughts as they are. Yet asking people questions and telling them things are problematic enterprises at the best of times, whether in the classroom or outside it. Problems arise particularly because:

- when *asking* questions, it is tempting to assume that the responder has taken the question to heart and that the reply represents what the responder thinks;
- when *telling* people things, it is tempting to assume that they then know what they have been told.

However, simply because it is easy to make inappropriate assumptions, it does not mean that asking or telling people is necessarily ineffective. When people are ready to hear what is said, telling can be very helpful—just as when a seed is sown in prepared ground. When people are engaged and interested, asking can, indeed, be stimulating—just as when a seed germinates and begins the cycle of growth again.

Despite the simplification necessary to think about asking and telling, it is important to bear in mind the complexity of classroom utterances. For instance, things may be said in order to control different aspects of a lesson:

- perhaps to re-focus attention after what, to the teacher, has been a diversion, as in 'So why do we use brackets?';
- perhaps to control the behaviour of an individual, as in 'What did I just say, Shona?';
- perhaps to signal the level of discussion intended, as in the difference between 'How wide is this desk?' and 'How can we find out how wide this desk is?', or between 'What is the name of a shape with three edges?' and 'What names shall we give to the shapes that we've found?'.

Sometimes a question hides an assertion, as in:

'What *have* you done?', meaning 'You have done something wrong'.

On the other hand, an assertion is sometimes actually a question, as in:

'You know something else', meaning 'There is something you are overlooking that I can see. What else do you know?'

There is also what Ainley (1987) has called the 'hovering' question, as in:

'This four-sided shape is called a …', or 'This ruler is for measuring in …'.

In the last case, learners are expected to fill in the missing word or words.

It might seem crucial that in a mathematics lesson the teacher's attention is focused on the correctness of the learners' responses. But if teachers focus their attention on getting correct responses, they are likely to engender an atmosphere of 'trying to get the answer the teacher wants' rather than create an atmosphere in which learners make conjectures and use their natural powers to take on new ideas. Similarly, a need for learners 'to understand' (Tahta, 1991) can misdirect teachers' attention and energy. Teachers may feel gratified that learners need help, but many interventions will drive learners closer to dependency and away from learning. The really well-aimed and well-timed intervention can liberate learners from being stuck in a rut, whereas over-used interventions which maintain teacher control can have the opposite effect.

Teachers may have multiple intentions when they speak. What they say can arise from a local, detailed concern about the mathematical ideas being presented, as well as from larger concerns about the conduct of the lesson and from global concerns about the nature of education and the discipline of mathematics. From a teacher's comments and questions, learners may form a view of the nature of mathematics and how it is to be pursued, even though teachers may be unaware of how what they say is being interpreted and may have entirely contrary beliefs to those that the learners assume. Learners may wrongly interpret a question by the teacher as intended to control them, or conversely, they may not notice a degree of control which enables lessons to proceed smoothly and efficiently.

Once the teacher has embarked on a lesson, asking, telling and listening are the central roles played by the teacher in the life of the learners. Although on the surface what distinguishes asking, telling and listening is what the teacher and learners are doing, it is useful to probe more deeply into what the teacher (and hence the learners) are actually attending to. For instance:

- when teachers are trying to reveal what learners are thinking, or to direct learners' attention, they find themselves *asking* questions;
- when teachers are attending to their own thoughts, they are likely to be caught up in *telling*;
- when teachers are attending to what the learners are thinking, they tend to be *listening*.

Asking questions and telling people things can, indeed, be productive, but only when the situation is appropriately structured. What this might mean is the subject of the rest of this chapter.

5.2   ASKING AND PROMPTING

Despite being asked questions by teachers throughout their time at school, not all learners appreciate being questioned, as the following quotes from learners suggest:

> 'I hated being asked questions and looking stupid if I didn't know.'
> 'I tried really hard but I just couldn't get the questions right. It wasn't worth trying.'

But when is asking really asking?

Researchers have delineated dozens of different forms of questions. In this section, three kinds of questions asked by teachers are considered: focusing questions, rehearsing questions and enquiring questions. These three categories provide an informative framework for further investigations and are based closely on distinctions proposed by Ainley (1987). Each category takes into account not just the words uttered, but also the intentions of the teacher. Those intentions may only be revealed when a learner responds and the teacher then reacts to that response.

This section then turns to questions and prompts that promote mathematical thinking, and embody the ways in which mathematicians think and act.

FOCUSING QUESTIONS

In the midst of a lesson, the teacher may 'see' something happening. For example:

- a learner has generated a sequence of numbers such as 1, 4, 9, ... but does not seem to see the pattern of square numbers (which leaps off the page at the teacher);
- learners are measuring their heights and struggling to articulate and record their results, but a link back to earlier work on naming decimals is clear to the teacher;
- learners are lost in the details of measuring and cannot see what calculations they need to do, yet to the teacher it is entirely obvious;
- learners are trying to refer to different parts of a diagram, and the teacher sees that they would be helped if they generated a notation or labelling system to make it easier.

The teacher is aware of something that would probably help the learners if they too were aware of it. When the teacher notices this possibility, it generates a desire to mention it. But is telling the best way to direct attention, or an outcome of what Mary Boole called "teacher lust": the desire to tell.

What the teacher says usually emerges as a question rather than as an instruction or a suggestion. Such questions, *focusing questions*, are designed to focus learners' attention on certain features that will help them to progress.

Teachers are often drawn into a sequence of focusing questions when the first one that they ask evokes no awareness in the learners. Take the situation where the teacher sees a pattern in the number sequence but the learner does not seem to see it, as in the sequence of square numbers 1, 4, 9 ... mentioned above. After each question the learner offers little or no response, so another question is asked:

1 'Can you see a pattern?'
2 'What is the same about this and this and this? What is different, or how do they differ?' (The teacher points to individual terms.)
3 'What connections are there between this and this and this?' (The teacher again points to the terms.)
4 (The teacher writes down the differences, 3, 5, 7..., between the given numbers 1, 4, 9... .) 'Do you recognise this sequence of numbers?'
5 'Will these numbers keep going up in odd numbers? Will you always get square numbers?'

Bauersfeld (1988) called this sort of questioning sequence the *funnel effect*. Each time the learner responds hesitantly, or not at all, the teacher feels drawn to be more precise, to ask something that the learner will surely be able to answer. Holt (1964, pp. 24–25) described an incident in which he eventually realised that one learner had a strategy of saying almost nothing until Holt had refined and simplified the question to the point where there was no risk in answering. Such situations arise when the teacher has a sharp sense of what they want the learner to know/see/say, and get sucked down the funnel by 'guess what's in my mind' questions. Familiarity plays a role in funnelling: a pattern of routine questions, asked without much real thought, can create a sterile atmosphere which forces the teacher into more and more specific questions.

Funnelling questions in themselves are neither good nor bad: rather their value depends on what the teacher and the learners think is happening. If the teacher believes something valuable is happening, or if the learners wait for the questions to become more detailed and precise without working at the more diffuse ones, then the interaction is likely to be a waste of time. If, on the other hand, the learners work hard at trying to relate ideas, then the funnelling may help to focus their attention in some unexpected or unfamiliar way, and the interaction could be fruitful. Overall, funnelling questions all have the same purpose, namely to get learners to see what the teacher sees; they do this by means of focusing the learners' attention.

The value of the term *funnelling* lies in the way that the teacher can bring it to mind in the midst of an interaction and so can become aware of being driven by 'guess what's in my mind'. This gives the teacher an opportunity to pause, and to choose to continue or to back off or to indicate more directly to learners the point of focus. The teacher can take alternative action provided that they have some alternative questions or prompts to use, such as 'Say the sequence to yourself in your head, or even out loud'.

Focusing questions often arise in interactions to which Vygotsky's notion of the zone of proximal development applies. For instance, when learners are working on a challenging task, the teacher may support their efforts by keeping track of the major goals so that the learners can attend to the details. At various points, the teacher may want to pull learners out of the details and to focus their attention on some global aspect. This can be done by using questions such as 'What was the question?', or 'What are you trying to do now?', or 'Does that calculation seem to be giving the sort of answer you expect?' A teacher might also use this kind of question to find out what a learner is doing and so help both teacher and learner to refocus.

Moreover, such questions serve to illustrate the way in which an internal mathematical monitor might work: in the midst of a calculation, experts may find themselves asking 'Is this the right calculation?', because their internal monitors are relatively well developed. By way of contrast, novices may be so deeply embedded in a calculation that they keep on struggling, getting deeper and deeper into a mire, without any internal monitor to make them pause and step back from what they are doing, and thereby get an overall picture. Vygotsky was convinced that the only way such a monitor will develop is through social interactions with others whose monitors are already active.

Focusing is a two-way activity. Teachers, as experts, are likely to be aware of connections, details and processes that the learners do not see, but learners are often aware of other features of questions that, due to familiarity, have become hidden to the teacher. It behoves a teacher to be sensitive to the learners' focus of attention and to try to adopt that focus some of the time.

REHEARSING QUESTIONS

It is common in teaching to ask questions that test learners' knowledge, memory or awareness. For example:

> 'What is a six-sided shape called?'
> 'What do you have to remember when using a protractor?'
> 'How do you round a number to one decimal place?'

Here the teacher certainly knows the answers, but is trying to get the learners to *rehearse* the answers for themselves. In *Maths Talk* (Mathematical Association, 1987), this is described as *checking up* and is contrasted with *exploratory* questioning. It may be enough that learners rehearse the words in their heads, though usually teachers like someone to give an answer so that everyone gets confirmation. Unfortunately, this often has the effect that only the person answering pays much attention to the answer! At the start of a lesson, when a teacher asks questions 'just to rehearse some of the ideas from last time', it can easily result in a sequence of laconic, even sullen, responses from the learners.

A large proportion of teacher questions are of the 'going over old ground' type. Rehearsal of terms and techniques is one good way to internalise ideas and begin to employ those ideas to express thoughts; this is far more effective than simply repeating memorised phrases.

Ainley (1987) pointed out that asking questions to which one already knows the answer is a cultural phenomenon associated with the European idea of school teaching. Aboriginal children in Australia come from a culture in which it is considered ill-mannered to ask a direct question, so to ask a question when one already knows the answer is thought to be bizarre! Constant questioning and extending the limits of knowledge of younger people may be typical of middle-class interaction patterns rather than endemic to society as a whole.

And yet it is hard to imagine not asking questions that stimulate learners to rehearse and reconstruct what they know. Where learners perceive questions as testing rather than as opportunities for rehearsing or for 'explaining to themselves' out loud in the presence of others, they may be confined by concerns about getting things right rather than being open to opportunities to get things *almost* right and then to modify and correct their answers.

The widespread popularity of *Who Wants to be a Millionaire?*, *Mastermind* and similar quizzes attests to the challenge that some people get from a series of probing questions: it is an opportunity for them to rehearse what they know, in public. Similarly challenging conditions can be set up in the classrooms; however, testing can gain the upper hand and start to drive teaching, so the teacher and hence the learners focus on what will be in the test, not on encountering the subject matter.

ENQUIRING QUESTIONS

Another form of questioning is sometimes called 'genuine', 'true' or 'authentic' questioning because it genuinely seeks information. Some examples of this form are:

'What were you thinking of when you wrote that down?' (In reference
to a line in a piece of learner's work.)
'How did you get that?'
'Can you put something on the container lid so that you can remember
how many beans there are in the container?'

When the teacher is seeking information in this way, the question can also
be labelled as *enquiring*. To stimulate learners to enquire for themselves, it
is useful, and possibly essential, to provide them with a model of enquiring
behaviour. It is not enough for the teacher to feign interest, because learners
will soon see through any pretence. Although it is not always easy for a
teacher to remain in a state of genuine enquiry about the mathematical content of their learners' work, it is possible to be genuinely interested in how
the learners are thinking, in how they express themselves, and in what they
'see' in their heads. Teachers who are evidently interested in their learners
are more likely to generate and support an enquiring attitude in them.

The phrase 'genuine question' is somewhat unfortunate. Firstly, because in
a classroom, questions are part of a power structure in which the teacher's
aim is to evaluate the learner's understanding so as to be of further assistance.
In addition, labelling a question as 'genuine' implies that other forms of
question are fake, misleading or of lesser value. The word 'genuine' is sometimes used as a value judgement, implying that all questions should or could
be 'genuine'. However, a question is not rendered fake or any less genuine
simply because it is deployed for rehearsal or focusing purposes.

All three forms of questioning—focusing, rehearsing and enquiring—have
their place in classroom practice. A particular form of questioning is problematic only if the teacher and the learners are unclear about what sort of
questioning is taking place. If learners are in the habit of interpreting all
teacher questions as *testing*, they must be shown that there are other types
of questioning, and that these present other opportunities.

LEARNING FROM ASKING

One of the difficulties with questions is that their meaning depends as much
on the teacher's intention as it does on the words used. Thus, most questions
that are intended as genuine enquiries can also be interpreted as focusing or
rehearsing questions or as some form of classroom control. For example, if
a teacher asks a learner:

'What do you see in the picture?' or 'What is the same and what is different about ...?'

the intention could be a genuine enquiry, meaning:

'I am interested in what you see and I cannot know unless you tell me.'

But it could equally well be interpreted as:

'There is something you should be able to spot, and your job is to work out what it is and to tell me.'

Or it could be seen as an attempt to focus attention without being directive:

'I have something in mind that I see. Do you see it too?'

or even to test:

'You are supposed to be attuned to seeing what I am seeing. Are you?'

In all cases, it is the follow-up reactions that signal both to teacher and learner what sort of an interaction is intended. One clue to a teacher's intention can be the amount of time before the teacher's next intervention. If learners are not given time to ponder and conjecture, then they will not be encouraged to reflect. On the other hand, when the teacher keeps silent, then the learners' attention can be focused on the content of the question rather than on just its form.

Often the teacher's intention in asking a question emerges only in reaction to the learner's responses, as the four following examples illustrate:

- The teacher notices a pattern of bricks in a wall and asks the learners 'What do you notice about the wall?' One learner replies with an observation about the colours of the bricks, another about the height, and yet another about the age of the wall. The teacher's reaction 'What about the pattern?' or 'How many bricks do you think there are?' reveals the existence of a definite focus in the original question.
- A learner is measuring with a ruler, but using the '1' mark as the starting point. The teacher asks 'What must you do when using a ruler?', and the learner replies 'Keep it still'. The teacher's reaction will reveal to both teacher and learner what sort of a question is being asked.
- In a secondary classroom, a task is set to work out all the lengths of the sides of a triangle and all the sizes of its angles, given just some of them. The data have been accumulated on a diagram, and the teacher, seeing that the law of cosines is the appropriate one to use, asks 'Which is it to be, sines or cosines?' Learners offer 'Sines', 'Cosines' and 'Measure them!', and the teacher's reaction to these suggestions reveals the specific intentions in asking the initial question.
- In a primary classroom, learners are playing a domino-type game with 'Logiblocks', in which each new block added to the sequence has to match the last block in exactly two attributes. The teacher sees a learner add an

inappropriate block to the chain and asks 'What are you matching for?' The learner may offer things like 'Colour', 'Colour and shape' or 'It looks right'. The teacher's reaction then reveals to the learner the original intention of the question.

In each of these four examples, the actual response summons up a reaction, and it is this reaction which is most informative about the original intention behind the question. Similarly, when a teacher asks a class a question that calls for an immediate response, it may only be when the teacher hears the response that they realise what kind of answer they was expecting.

The response also tells the teacher how the learners have interpreted the question. Learners will either interpret a question so that it corresponds to something they can do, or else balk and do nothing until the teacher provides further help. Of course, it is entirely natural to respond to a question by seizing upon the first thing that comes to mind. Sometimes this leads to trouble, and therefore the teacher will want to encourage learners to 'think before they leap'. Sometimes the simplicity of the first thing that comes to mind leads to a simple approach that works in the given context but perhaps is more difficult or impossible when the task is harder. This is one reason for not always starting with simple examples, but instead sometimes starting with complex examples that can then be simplified in front of the class, as discussed in Chapter 3.

Asking questions such as 'How might you go about this calculation?' or 'How many different ways could you do this calculation?' can reveal those learners who are unaware that they have a choice. For example, learners asked for the 'best operation' to count the number of eggs in ten boxes, each containing six eggs, may respond by using adding or counting rather than with multiplication, because those operations are seen as 'best for them'. Vygotsky (1934/1986) suggested that children and adults see the same object differently: '... the child's framework is purely situational with the word tied to something concrete, while the adult's framework is conceptual' (p. 133).

A wider range of questions and prompts that teachers might use are now examined.

QUESTIONS AND PROMPTS

Teachers do not only ask questions of their learners. They also use a repertoire of prompts, instructions and reminders to stimulate activity. The kinds of questions and prompts that are presented to learners will influence their view of mathematics. If the majority of prompts are of the form 'Do this calculation' or 'Can anyone tell me how to do this?', then learners are encouraged to think that the most important feature of mathematics is to get the correct answer. This results in a severely limited view of mathematics, in which there is no room for creativity or originality. Although many learners then come to feel secure in a subject where there are fixed methods and out-comes, they are consequently unable to cope when they are required to think creatively or make connections between ideas.

Mathematicians have a much greater range of types of questions that they ask themselves, questions such as 'Can I find a special case of this result?' or 'Is that always true?' Following Dyrszlag (1984), Watson and Mason (1998) suggested that if learners are to gain a rounded view of mathematics, they need to be immersed in similar types of question.

Watson and Mason (1998, p.7) listed words that denote processes or actions that mathematicians employ when they pose and tackle mathematical problems:

exemplifying	comparing	reversing	justifying
specialising	sorting	altering	verifying
completing	organising	generalising	convincing
deleting	changing	conjecturing	refuting
correcting	varying	explaining	

These words are associated with mathematical thinking, reasoning and the understanding of a concept, rather than words such as calculating, solving, drawing and measuring which are concerned with techniques.

Drawing on Dyrszlag's writing, Watson and Mason grouped the processes referred to in the list above and gave typical questions and prompts associated with each group. (Please see the table overleaf.)

Any one of these categories can be applied to various mathematical 'objects', such as those in the following list:

definitions	conjectures and problems
facts, theorems and properties	representation, notation and symbolisation
examples and counter-examples	explanations, justifications, proofs and reasoning
techniques and instructions	links, relationships and connections

For instance, when applying the 'Changing, varying, reversing, altering' category to definitions, the teacher might say 'Change the definition of a square so that it describes a rectangle'. Applying questions from the same category to a technique might result in an instruction like 'Multiply 23 by 17 in at least two ways'.

Exemplifying Specialising	Give me one or more examples of … . Describe (show, choose, draw, find, …) an example of … . Is … an example of …? What makes … an example? Find a counter-example of … . Are there any special examples of …?
Completing Deleting Correcting	What *must* be $\left\{ \begin{array}{l} \text{added} \\ \text{removed} \\ \text{altered} \end{array} \right\}$ in order to $\left\{ \begin{array}{l} \text{allow} \\ \text{ensure} \\ \text{contradict} \end{array} \right\}$ …? What *can* be $\left\{ \begin{array}{l} \text{added} \\ \text{removed} \\ \text{altered} \end{array} \right\}$ without affecting …? What needs to be changed so that …? Tell me what is wrong with … .
Comparing Sorting Organising	What is the same and what is different about …? Is it or is it not …? Sort or organise the following according to … .
Changing Varying Reversing Altering	Change … in response to imposed constraints. What if …? Do … in two (or more) ways. Which is quickest, easiest, …? If this is the answer to a similar question, what was the question? Alter an aspect of something to see the required effect.
Generalising Conjecturing	What happens in general? Of what is this a special case? Is it always, sometimes, never …? Describe all possible … as succinctly as you can. What can change and what has to stay the same so that … is still true?
Explaining Justifying Verifying Convincing Refuting	Explain why … . How is … used in …? Explain the role or use of … . Give a reason … (using or not using …). How can you be sure that …? Convince me that … . Tell me what is wrong with … . Is it ever false that …? (Always true that …?)

Of course, this approach is not the equivalent of a mechanical device that will automatically produce mathematical thinking in learners. The lists are of no use to teachers until they have been linked to the teachers' own experience, and integrated into their classroom practice. Only then can teachers use them to help learners to experience aspects of mathematical thinking.

5.3   TELLING, DISCUSSING AND LISTENING

WHEN TELLING IS REALLY 'TELLING'

> If you scale a shape by a factor of k, then the perimeter is scaled by a factor of k and the area is scaled by a factor of k^2.

This statement could be part of a formal lecture; it could also signal a conjecture to be investigated either at a learner's suggestion, or due to the teacher's repetition of a learner's or teacher's conjecture; it could also be a summary of findings at the end of a period of work. In other words, how a statement is interpreted depends on its context and on the listener's expectations, just as with questions.

In general terms, the statement could be seen as an example of *telling*. As a mode of interacting with learners, telling received a particularly bad press in the 1980s. Emphasis on investigation, group work and learner experience led to a dislike of and reaction against exposition. This hostility was probably prompted by memories of sitting in lessons and being told things endlessly by a succession of teachers. However, in 2001, HMI found that in mathematics lessons, 70% of the time involved telling. When telling is used as a device to transfer pre-packaged knowledge and skill from one person to another, or to impose it on learners, it deserves reproach. Yet there are many circumstances in which telling is not only proper and effective, but essential for giving information. Care is needed to try to make sure that learners are ready to take on board the information and are in a position to reconstruct for themselves what they are being told. Cobb (1988) identified 'teaching by imposition' at one end of a spectrum with 'teaching by negotiation' at the other end. Somewhere close to the middle can be found 'teaching through telling'; this requires learners to be ready to listen. They learn to listen through enculturation into mathematical practices and into ways of working that involve negotiation and discussion.

People can be told things in more than one way. Indeed, Ainley (1987) suggested that not only assertions but also questions do their share of telling people things—things about the speaker's interests and concerns, as well as things about how the subsequent lesson is likely to develop. People tell each other things all the time, often very effectively, though frequently the significant communication is not the literal meaning of the words, but the exchange of attention—what Berne (1964) called 'stroking'. Examples of stroking include ritualised exchanges, such as 'Good morning', 'How are you?' and 'Thank you', which serve to lubricate the cogs of personal interactions. Communication is more difficult when such stroking exchanges are absent.

There are social settings in which telling is a response to an implicit or explicit request. For example:

- to find out the bus or train times or when a TV programme is going to be broadcast, information is wanted, not a series of questions or deflections;
- to find out how to use some new device, an invitation to 'explore for yourself' as the sole contents of the instruction booklet could be dangerous as well as unhelpful;
- to find out what someone else thinks about an issue, it is not effective to get that person to *ask* a series of questions;
- to hear what someone has to say in a lecture, it is no good to be put into small groups for the whole session to discuss what participants think—choosing to attend a lecture suggests a wish to be told.

In school contexts, where teachers are expected to do something that causes learners to learn, it is often not clear whether telling or not telling is likely to be more effective. In the midst of a complicated calculation when the answer to 'seven eights' is needed but momentarily forgotten, the answer is what is wanted, not an invitation to work it out. But is that the most fruitful thing for the learner? There are, of course, no rules. Any principles, such as 'Never tell them something they can work out for themselves' or 'Treat them as adults—if they ask a direct question, give them the answer', over-stress one aspect at the expense of another.

A common feature of social situations in which telling is desirable is that someone wants to know something. Unfortunately, this is not always the case in the classroom. Learners do not necessarily want to know things: sometimes they just want to get by and avoid being asked questions. However, telling may stimulate interest and engage learners in activity that generates surprise or contradiction. As a result, learners may want to learn and may begin to use their initiative.

It is natural to want to tell people things and natural to want to be told things. Since it is impossible in one lifetime to discover everything for yourself, people try to learn from others. There are two difficulties about 'telling' things to other people. One, which has already been mentioned, is the assumption that the hearers know what they have been told. There is plenty of evidence that this assumption is a common state of affairs between teachers and learners:

'If I've told you once, I've told you a hundred times!'
'You were told that last (week, month, term, year, ...).'

Learners collude in this assumption, for they often say, or behave, as if they believe:

'Just tell me once more and then I'll understand.'

The other problem with telling is that people may be told things that they are just about to find out for themselves, and so telling disrupts their productive thought or deflates them by taking away the pleasure of discovery. Boole (Tahta, 1972, p. 11) drew attention to *teacher lust*: the teacher's desire to explain to others. In general, a desire to explain can be positive; for instance, when learners become so excited by their insight that they feel compelled to tell someone else. In this situation the person who learns most is the explainer, not the listener. However, teacher lust is a negative manifestation of the desire to explain: it arises when a teacher feels the urge to explain something to learners who are perched on the edge of coming to it for themselves. Hence the notion that the teacher is trying to do for the learners what they cannot yet do for themselves. This is another way of thinking about Scaffolding-and-Fading, as well as about Vygotsky's notion of the zone of proximal development (Chapter 4).

If it is not sensible to assume that people know what they have been told, why tell them at all? Yet, for teachers, to avoid telling would be just as perverse as to rely upon telling as the main form of interaction with learners. To refuse to tell on the grounds that it does not help (because 'they still don't know') contradicts the belief that telling is liable to deflate or disrupt. If telling is indeed so ineffective, then it would not be possible to disrupt or deflate people by telling!

Learning is a much more complex process than a simple cause-and-effect relationship between what the teacher says and what the learners learn. It makes sense to tell people things when they are receptive and are able to make connections with what is being said and can assimilate it, but are not yet able to work it out quickly for themselves. If people are not receptive to what is being said, then they are unlikely to make much of it. Take the case of the learner who remarked to his mother:

'Mr P is being very silly at the moment. He's doing all sorts of complicated things to add fractions together, when all he needs to do is add the tops and add the bottoms.'

This learner may not have been in a position to take on board the teacher's algorithm. Nor was the learner who was confident about using a ruler, but always started from the '1' mark 'because you always start counting from one'.

For telling to be successful, the ground needs to be adequately prepared by arousing listeners' interest or sharpening their awareness. There usually has to be something problematic or striking to engage learners if they are to be in a position to be told something effectively. What is said has then to relate to the learners' experience and to call upon their confidence, in order to push at the boundaries of their current understanding.

Movshovits-Hadar (1988) suggested that it is possible for teachers to invigorate learners by re-enthusing themselves and re-experiencing surprise or wonder at even the simplest of mathematical facts. By re-entering the state of mystery or excitement, or by reconsidering some problem or surprise that the ideas resolve, it is possible for teachers to awaken that same surprise in learners.

WHEN DISCUSSING REALLY HELPS

Since teachers ask most of the questions in lessons, it is understandable that learners usually respond to the teacher. The trouble with this is that many learners will attend closely to what the teacher says but will ignore what fellow learners say. In order to promote constructive dialogue so that learners can work productively together with or without the teacher, it is usually necessary to spend time training learners in listening and discussion.

Reports of lessons often include 'We discussed ...', but observation of such lessons may reveal that the teacher told the learners a few things and perhaps asked a few questions to which the learners gave, at most, succinct answers. This sort of 'discussion' is a travesty of what is actually possible.

MATHEMATICAL DISCUSSION

Can mathematics actually be discussed in the general sense of the term 'discussed'? How can you have opinions in mathematics? If discussion is seen as a mode of interaction in which individuals try to express their ideas and, through expressing them, come to a better understanding, then mathematical discussion is, indeed, possible. It provides an opportunity for learners to distance themselves from their own ideas, to become articulate about their insights, to examine their ideas and intuitions critically, and to make modifications supported by the suggestions of others. In the process, it releases the teacher from having to be present all the time; recent studies of discussion have often made use of a computer as a stimulus for discussion (see, for example, Healy and Hoyles, 2000). So learning *can* take place without the teacher.

In order to accomplish this distancing, learners need to struggle to express their own thinking and also need to listen to and learn from the thinking of others. They should become habituated to a conjecturing atmosphere. This requires careful development by the teacher, possibly by training learners to listen and ponder as well as to speak clearly enough for others to hear and make sense of their utterances. It also requires teachers to have confidence in what they are doing and to instil confidence in the learners.

Until learners are experienced in discussion techniques and are willing to open up and offer their ideas, it is difficult to generate discussion simply by asking questions. It is useful to have some object or situation or assertion for learners to focus on, rather than just embarking on a free-flowing discussion without having considered in advance the ideas that are likely to arise. It is essential for the teacher to be attuned to the language patterns that learners are likely to use and to be sensitive to the images and awarenesses that prompt learners to say what they do. If the teacher decides in advance where the discussion will lead, there is a risk of it becoming merely a question-and-answer session. There is a delicate balance to be achieved between preparation and spontaneity.

Pirie and Schwarzenberger (1988, p. 461) set out a list of possible requirements for a mathematical discussion:

- *It must be purposeful talk.* That is, there must be well-defined goals even if not every participant is aware of them. These goals may have been set by the group or by the teacher, but they must be, implicitly or explicitly, accepted by the group as a whole.
- *It must be on a mathematical subject.* That is, the goals themselves, or else subsidiary goals that emerge during the course of the talking, must be expressed in terms of mathematical content or process.
- *There must be genuine learner contributions.* That is, there must be input from at least some of the learners, which helps to move the discussion or thinking forward. The emphasis here is on the introduction of new elements to the discussion, rather than on mere passive responses such as factual answers to teachers' questions.
- *There must be interaction.* That is, there must be indications that the movement within the talk has been picked up by other participants. This may be evidenced by changes of attitude within the group, by linguistic clues of mental acknowledgement, or by physical reactions which show that critical listening has taken place, but *not* by mere instrumental reaction to being told what to do by the teacher or by another learner.

Bartolini-Bussi (1990), in working with mathematics teachers over a three-year period, identified three types of mathematical discussion: discussion *before*, *within* and *after* work on a topic.

- *Before:* Usually short discussions that involve learners in discussing how they see the task or question which has been presented. Such discussions can involve the teacher in collecting together learners' ideas and associations with a topic before the learners get down to a pre-arranged task. For example, the teacher might ask learners what they know about π or about cutting things into portions. The purpose here is to remind learners of what they already know in relation to the current topic.

- *Within:* Discussions that arise between learners as they work on a task or explore a suggestion. It is not easy for a teacher to manage this type of discussion, because it is spontaneous and because the teacher often finds it difficult to see what aspects are important and what aspects are peripheral to learners' understanding. However, such discussions play a significant social role in the individual learner's construction of meaning. For example, it is often the case that a learner who says little understands a lot, while a learner who talks a lot may not be listening or even working very hard at making sense of things. A learner who seems to be using the words in incorrect or misguided ways may actually be sorting out what they mean, while another who uses the words correctly may be repeating them from memory rather than employing them to express ideas. With repeating from memory, the key thing is that, in becoming familiar with using technical terms and stock phrases, those terms grow in meaning for the learner. Compare this with scaffolding where the key thing is the removal of the scaffolding—the *fading.*

- *After:* Discussions that act as an impetus to reflect and try to relate ideas so as to form some sort of 'story' summarising what the topic is about. In lessons, time is often seen as an enemy and tiredness takes over, so reflection is omitted. Yet if the teacher is not seen to value such activity, learners will also tend to devalue it, and the whole point of a series of lessons may be largely wasted.

Bartolini-Bussi also suggested that the teacher has two major roles to play in a discussion (and therefore in training the learners for when the teacher is *not* present), namely as *moderator* of the interactions and as *mediator* of the mathematical ideas. As moderator, the teacher makes sure that the learners are listening, keeps the focus on the learners by deflecting attention away from the teacher and encourages learners to listen to each other and to question and modify what is said. As mediator, the teacher is available to help with technical terms and to introduce examples which may require more thought but which the learners may not find for themselves.

It might be expected that when learners discuss ideas among themselves, this would throw up disagreements and conflicts which would, in turn, lead to resolution and hence learning, with or without recourse to the teacher. But it is clear that discussion itself, being a social interaction, will not necessarily reveal disagreements and, even if it does, may not provide any support for reconciling them. Furthermore, learners may confirm mistakes rather than correct them—or may even develop new ones. Must the teacher be present for correct learning to occur? Certainly there is more to it than simply putting learners in groups and getting them to discuss (see, for example, Brissenden, 1980).

Bell and colleagues (Bell, 1986; Bell, Brekke and Swan, 1987a, b; Bell and Purdy, 1986) proposed that one fruitful way for a teacher to generate discussion is to set tasks that reverse the usual kind of exercise questions which learners are given when they are practising a skill. So, instead of 'What is 2×3?', ask 'What question would give 6 as an answer?' Instead of 'What is $\frac{2}{3} + \frac{4}{5}$?', tell learners that someone got the answer $\frac{6}{8}$ but that it was marked wrong, and ask them what they think the learner did and why was it incorrect. Instead of asking learners to put 0.234, 0.56 and 0.7 in order of increasing size, tell them that when learners are asked to do this, some say 0.234 is the largest, even though it is smaller than 0.56, and then ask why those learners might do this.

In each case, the learners' attention is being drawn to a difficulty that they themselves might have. Offering this difficulty for discussion and giving learners a chance to explain how they think about it encourages the learners to expose and distance themselves from erroneous or incomplete thinking; this makes it more likely that they can correct their own thinking.

Mathematical discussion involves learners taking opportunities to express to the teacher and to each other what they are seeing and thinking. It is not the mathematical facts that are 'up for discussion' in the sense of negotiation, but rather the learners' thinking—their constructions and ideas that embed those facts and give them meaning. Discussion is a way to support learners, individually and collectively, in developing confidence in mathematical ideas and the language used to express those ideas.

WHEN LISTENING IS REALLY TEACHING

> I have become a better listener. Teachers are basically talkers who feel a strong desire to share their knowledge with other people. Children are no different. If we really make an effort to listen to our students, we will become the richer for it.
>
> Teacher comment after a year as a participant in a research programme Quoted in Cobb, Wood and Yackel, 1990, p. 135.

The frameworks of Manipulating–Getting-a-sense-of–Articulating, Doing–Talking–Recording and Scaffolding-and-Fading all suggest that learning is enhanced when learners talk about their thoughts as part of becoming more articulate and before making succinct records. It follows that an important dimension of teaching is listening. The learners benefit from struggling to express themselves, and from modifying what they say and think on the basis of what they hear themselves and others say. A constant diet of expert talk by a teacher will dull learners' urge to try to articulate their thoughts, and will lead them to expect learning to take place as a result of just listening.

Belenky and colleagues (Belenky *et al.*, 1986), in a study that examined differences between the ways that men and women interact in educational settings, identified major features of interactions. They observed that to see clearly requires the viewer to stand back and get an overview, and that to hear distinctly requires not only proximity but also a letting-go of personal concerns. They also noted that seeing suggests a single actor, whereas speaking and listening imply the possibility of dialogue amongst several people.

Listening is different from hearing. You can *hear* the radio while you prepare a meal, but when you *listen* to the radio you sometimes forget what you are doing. When you are doodling in a meeting, your attention can swing from listening to hearing and back again. Listening is a form of focused attention, just as looking is a focused version of seeing.

A teacher can sometimes tell whether learners are listening by subtle changes in their posture, gestures or voice tones. These indicators may be those that are typical of concentration (slight leaning forward, perhaps, and a certain stillness) but may also involve rapid movement and interruptions. When teachers circulate round the classroom while learners are working, they often cajole learners, tell them things and direct their attention. But it can be effective to circulate round the classroom listening—not merely hearing what is said, but actively listening, paying attention to learners' postures, gestures and tones of voice.

It is a paradox of teaching that learners often learn when teachers are listening to them rather than the other way round. This could be explained by claiming that through listening, teachers discover what further interventions might be helpful. However, it may be the stimulus of interested and caring attention from a listening teacher that encourages the learners to articulate their fuzzy ideas and so enables them to integrate their experiences. Freudenthal (1991, p. 55) described the 'teaching by listening' paradox as a tension between 'the force of teaching and the freedom of learning'.

A *listening* teacher can connect two assertions by different people, even when those assertions are incompatible and therefore need negotiating. A *talking* teacher, on the other hand, would need to launch into an exposition in order to clarify the ideas—but for whom? A listening teacher can focus learners' attention by the movement of an eyebrow or the jab of a finger, whereas a talking teacher would need to intervene verbally and re-establish personal control of the situation.

Teaching through listening is a subtle matter, not amenable to hard and fast rules. Of course, listening teachers also talk, but when they talk, what they say is more likely to be pertinent and heeded by listeners. Listening leads to genuine questions ('How did you get that?', 'Will that work in other situations?') which inform the teacher about the thinking processes of the learners. Such questions lead to something worth listening to.

Davis (1996, pp. 47–50) drew attention to differences between *listening for* (what you are expecting), *listening to* (what is being said) and *listening in order to* (contradict, add to, make use of what is said).

Listening for is listening with an outcome in mind. It has its place, but is not as significant in the classroom as listening to, which involves quite a different form of attention.

You can hear someone speaking, and you can *listen to* what is being said as a sequence of words, but you can also *listen to* something that encompasses both the person and the words, and so gain access to a world of experience. Listening *to* is an active act of attention.

Learners can often detect whether or not they are being listened to. Being truly listened to can be inspiring, while being listened to peripherally can be disheartening. Perhaps the most important aspect for learners of being listened to is that they participate in an action. They try to articulate the jumble of thoughts and uncertainties that tumble around in their minds. This is a vital part of the complex action that is called learning. As the old adage goes, 'You only really learn when you try to teach'. By using newly encountered symbols and technical terms to express yourself, you integrate them into your functioning and, at the same time, enhance your ability to notice in the future. Likewise, expressing to yourself and to colleagues the insights into your own experience triggered by this course enhances the possibility that your future actions will be more richly informed.

Davis's third type of listening, *listening in order to*, can become, at its best, a conversation. A conversation is more than a sequence of utterances or assertions. There is a genuine meeting of minds as two or more people engage with each other, mediated by a common topic or issue. A conversation can assist the transformation of awareness that is associated with learning. Gadamer (1990) mirrored Vygotsky's general notion of cultural tools as mediators, by pointing to the way in which, in an effective conversation, the topic is what brings the participants together, but that each participant learns by attending to what is said. The interaction is even mirrored in the participants' physical movements, now leaning forward, now sitting back, now agitated, now calm.

Davis drew upon the philosophers Merleau-Ponty and Gadamer to stress the uniting aspects of conversation:

> We are thus *joined* in the conversation, ... [we are] in conversational unity, we become capable of greater insight and deeper understanding, capable even of cutting beneath the conscious intent of the speaker.
>
> Davis, 1996, p. 41.

It is difficult to over-stress the importance of listening in teaching. As Davis said:

> Occurring somewhere between the surety of the known and the uncertainty of the unknown, the act of listening is similar to the project of education. It is, after all, when we are not certain that we are compelled to listen. Our listening is always and already in the transformative space of learning.
>
> Davis, 1996, p. xxiv.

A teacher who creates conditions in which learners feel the urge to express their emerging conjectures can achieve a great deal of teaching by listening to learners. When listening is followed by a well-placed remark from the teacher ('Have you thought about ...?', 'What did you do last week in a similar situation?'), learners can accomplish more than they could on their own, and the teacher provides a shared experience that the learners can build upon during reflection and in the midst of further activity in the future.

SUMMARY

Much has been said in this chapter about interactions that sustain activity, and much more could be said. It has been suggested that asking questions can be productive, but only if it is clear what kind of questions they are and provided they are as mathematically based as possible. It has also been suggested that telling things to people is entirely natural, and that it can be effective but only if those hearing are in a position to listen, perhaps by having engaged in some relevant activity themselves.

Asking, telling and listening are three basic actions in which both teacher and learner can participate, but each has subtle variations. An apparent question can actually be an instruction, and people can apparently be being told something when it is actually a question. Asking questions can be productive, but it can also be a cover for reinforcing control and dependency; telling can be effective, but also ineffective; listening can be passive or active.

In mathematics lessons, teachers, learners and mathematics are intertwined in different modes, all within the milieu of the classroom. Exposition can be effective but is hard to break out of, while explanation based on listening is hard to remain in. Making balanced use of all modes is more effective than the restricted use of just one or two of them.

6
PROGRESSION IN MATHEMATICS

In previous chapters, the focus has shifted from mathematical topics, to tasks, to activity and finally to interaction. In this chapter, we look back by considering the notion of *progression*—progression for individuals within a lesson, over several lessons and over the longer term.

The term *progression* is unfortunate in that it summons up an image of a staircase of levels, with learners advancing from step to step. But learning is neither uniform nor uni-directional. Sometimes learners realise that they did not understand what they thought they understood. They try to tell someone how to carry out a procedure, and suddenly realise that there are gaps in their knowledge. Or they try to program a computer to carry out a calculation sequence, and find there are wrinkles in their understanding that they have been overlooking. Nevertheless, learners *do* develop: they are able to do things that they previously could not do, and they have new awarenesses.

From the point of view of learners, the most obvious form of progression is the way in which they advance through a scheme of work. This is, in part, because they are often unaware of the ways in which they have matured mathematically. For the teacher, however, it is very limiting to see progression in terms of completed tasks, not least because it is all too easy for learners to respond to tasks without necessarily contacting the mathematical ideas and thinking as intended. Moreover, a sequence of tasks may not actually develop learners' thinking if the level of challenge in the tasks is inappropriate.

Insufficiently challenging tasks require little thought from learners, while overly challenging ones mean that learners need a great deal of help. In neither case are learners likely to deepen their experience of the underlying mathematical ideas. All too often, textbook authors and publishers, fearful that learners (and perhaps teachers) will not know what to do, only provide tasks with little challenge. But if learners are not challenged, it is unlikely that any learning will take place.

The notion of progression in learning assumes that learning can continue to develop. But what is it that changes and develops? 'It' probably includes learners' competence and facility with techniques, their use of strategies, their awareness of mathematical themes and the use of their powers to think mathematically.

We now examine how some researchers think about the aspects of learners' mathematical understanding that are essential for progression.

6.1   UNDERSTANDING MATHEMATICS

Kilpatrick, Swafford and Findell (2001) produced a comprehensive, research-based analysis of the teaching and learning of mathematics. They characterised mathematical understanding as having five strands:

- conceptual understanding;
- procedural fluency;
- productive disposition;
- adaptive reasoning;
- strategic competence.

The first three of these strands can be related to awareness, behaviour and emotion.

Conceptual understanding involves the education of awareness through undertaking tasks and engaging in activity in order to bring pertinent actions to attention. Teachers, by reminding themselves of the various images and links that they want learners to incorporate in their concept-image, are in a position to choose topics and ways of interacting during learner activity that support the achievement of those aims.

Procedural fluency results from the training of behaviour. But the training of behaviour on its own produces lack of flexibility and lack of *adaptive reasoning*. Behaviour is more appropriately trained in the context of educating awareness. This means that tasks should be chosen which call upon learners to rehearse recently met ideas as a means of encountering and gaining experience of some new concept, rather than the focus always being on practice for the sake of perfecting practice.

The third strand, *productive disposition,* is perhaps the most unfamiliar; it refers to developing habits of identifying and tackling problems, taking initiatives to construct examples and counter-examples, and trying to justify mathematical conjectures.

As an alternative to Kilpatrick's approach, Marton (Marton and Booth, 1997), who coined the expression *dimensions-of-possible-variation* (see Chapter 1 above), argued that learning can be seen in terms of extending the awareness of dimensions-of-possible-variation associated with tasks, techniques, concepts and contexts, as well as the awareness of the range-of-permissible-change within each of those dimensions. To appreciate and be able to use a concept, learners have to know what is allowed to change so an object still meets the definition or criteria for that concept, what features of situations the concept is most usefully applied to, and what techniques are usually associated with it.

6.2   PROGRESSION WITHIN AND BETWEEN LESSONS

The frameworks discussed in Chapters 3 and 4, *See–Experience–Master, Doing–Talking–Recording, Manipulating–Getting-a-sense-of–Articulating* and *Scaffolding-and-Fading,* together provide a way of thinking about learning as movement. Learning advances as new experiences support the emergence of a sense of an idea by means of doing and talking, as well as by trying to record what is said and imagined, and trying to articulate succinctly ideas and insights that may initially be vague and fuzzy. When difficulties arise, learners revert to familiar objects and situations to try to see what is going on; as their confidence returns, they try to generalise and reconstruct the ideas for themselves.

Progression within a lesson and between lessons can be seen in terms of:

- encountering a new idea, topic, strategy or approach;
- further experience of important ideas, themes, powers, topics and/or terms;
- development of concept-images;
- more articulate awareness of the structure of a topic, the use of a technique and/or the range of situations in which a topic or technique is likely to arise.

Work on one topic can be designed to involve the use of ideas, strategies and techniques from earlier topics, sometimes linked together through the contexts in which they appear or by the mathematical themes which they exemplify.

Scaffolding-and-Fading provides a way for teachers to be aware of learners' progress. If a teacher is still asking the same kind of questions at the end of term as at the beginning, then 'fading' is unlikely to be happening and learners may be just as dependent on the teacher as they were at the outset.

However, if learners *are* taking up some of the kinds of questions that a teacher asks and using them for themselves, then there is evidence of progression in their awareness of how to learn mathematics: they are becoming more independent, and more efficient and effective in approaching new topics.

Evidence of progression in learners' mathematical thinking is also revealed by their increasing independence. While at the beginning of a topic, learners may depend on the teacher for useful examples, they show progress when, as they work through the topic, they begin to construct their own examples to illustrate relevant features or to test conjectures and provide counter-examples.

When learners are using their initiative, asking and formulating questions for themselves, helping each other when stuck rather than turning immediately to the teacher, and using their many natural powers in increasingly sophisticated ways, then progression is taking place.

6.3   PROGRESSION IN THE LONGER TERM

Progression in the longer term equates to learning, and although it is not easy to capture precisely what is meant by progression in this sense, several mathematics educators have produced plausible theories.

The idea of a *procept* was put forward by Gray and Tall (1994). This notion was developed to try to capture a process of change in learners' perceptions as the actions that they carry out without explicit awareness gradually become objects in themselves. Processes become concepts while, at the same time, 'theorems' and 'facts' which are implicit within the actions emerge as properties that can be discussed and justified.

For example, consider the development of learners' ideas of symmetry. Learners might first recognise symmetry in objects by becoming aware of objects being 'the same on both sides'. They may then use a mirror to see symmetry in shapes, with an essential property being that the mirror image coincides with the 'other half' of the shape. Subsequently they will be able to think of a coinciding mirror image as a property of objects generally and not just of specific ones that they have encountered. The notion of a 'mirror line' emerges and, with it, the action of forming a reflection. Eventually, a reflection in a particular mirror line is denoted by a letter and becomes an object which can itself be manipulated by combining it with other mirror reflections.

Sfard (1994) used the term *reification* (meaning 'becoming an object') to refer to the transformation in a learner's awareness, in going from actions to implicit actions to the action becoming a mathematical 'thing' that can be manipulated. Freudenthal (1983, 1991) used the term *condensation* to refer to something very similar, in which learners at first require many words and even pictures to describe something, but these are gradually condensed into a single symbol or a succinct definition or a handy label. For instance, 'angle' becomes a technical term which summarises not only a wealth of bodily experience with turning, but also a way of looking at the region around a point where two lines (or, indeed, two smooth curves) meet.

van Hiele (1986) studied the way in which reification or condensation comes about over time in the context of geometry. van Hiele described the development that takes place in a learner, who at first attends only to a whole shape, and then progresses to discerning parts of it; this is followed by seeing parts as an aspect or attribute of the whole shape, and then seeing relationships amongst those attributes. Finally, the learner can see certain relationships as properties that characterise the whole shape.

The example of symmetry considered above can fit into the process just outlined. At each 'level', what was previously the focus of attention becomes an object which is thought about and used at the next level, leading eventually to informal and formal reasoning. However, the connotation of a succession of levels over time is unfortunate: once a level has been acquired a learner does not use only the way of thinking of that level, but will operate on all of the levels, often simultaneously or in quick succession.

SUMMARY

Progression can be measured in terms of developing competence and fluency of techniques, but such an impoverished perspective is likely to overlook the growing sophistication of learners' thinking, as manifested by their use of their powers, by their ability to use their initiative, by the growth of their awareness of mathematical themes and thinking processes, as well as by their increasing ability to make connections between mathematical topics.

The idea of progression in understanding is hard to pin down. Several approaches that try to be more precise about its meaning have been discussed. These have been found useful by teachers wishing to sensitise themselves to learners' ups and downs in understanding, assimilating and accommodating the ideas, techniques and ways of speaking and thinking that characterise mathematics.

7
EPILOGUE

The heart of teaching is interaction with learners; the rest is preparation to make this interaction useful. Activity by learners creates opportunities for pedagogically effective interaction. Where learners take the initiative, the teaching can guide, support and foster learners' use of their mathematical powers to investigate phenomena which can be analysed mathematically.

Activity is initiated by tasks, which are constructed so as to put the learner in contact with important ideas in curriculum topics and major mathematical themes and, wherever possible, to enable learners to take the initiative. Progression involves the learners' increasingly sophisticated use of mathematical powers and themes, and increased sensitivity to opportunities for using those powers, associated techniques and ways of thinking, both within school contexts and beyond.

A number of frameworks have proved effective in informing teachers' preparation for teaching, the conduct of lessons and reflection upon these. These frameworks have been used in the construction of this book, some more overtly than others. Particularly salient has been the use of dimensions-of-possible-variation and the associated ranges-of-permissible-change to suggest how tasks, task-presentation, techniques, concepts and contexts can be 'psychologised', in the words of John Dewey: that is, how they can be worked on with and by learners in a pedagogically effective manner.

We hope that your reflection upon the ideas discussed in this book will help you to develop your own teaching and will enable you to design and present tasks so that the learners with whom you work will interact increasingly productively.

Bibliography

Ainley, J. (1987) 'Telling questions', *Mathematics Teaching*, no. 118, pp. 24–26.

Artigue, M. (1993) 'Didactical engineering as a framework for the conception of teaching products' in Biehler, R., Scholz, R., Strasser R. and Winkelman, B. (eds) *Didactics of Mathematics as a Scientific Discipline*, Dordrecht, Kluwer.

Askew, M., Brown, M., Rhodes, V., Johnson, D. and William D. (1997) *Effective Teachers of Numeracy*, London, Kings College.

Banwell, C., Tahta, D. and Saunders, K. (1972) (updated 1986) *Starting Points for Teaching Mathematics in Middle and Secondary Schools*, Diss, Tarquin.

Bartolini-Bussi, M. G. (1990) 'Learning situations and experiential domains relevant to early childhood mathematics' in Steffe, L. P. and Wood, T. (eds) *Transforming Children's Mathematics Education: international perspectives*, Hillsdale, Lawrence Erlbaum Associates.

Bauersfeld, H. (1980) 'Hidden dimensions in the so-called reality of a mathematics classroom', *Educational Studies in Mathematics*, no. 11, pp. 23–41.

Bauersfeld, H. (1988) 'Interaction, construction, and knowledge—alternative perspectives for mathematics education' in Grouws, D.A. and Cooney, T.A. (eds) *Perspectives on Research on Effective Mathematics Teaching: research agenda for mathematics education*, vol. 1, pp. 27–46, Reston, Virginia, NCTM and Lawrence Erlbaum Associates.

Becker, J. and Shimada, S. (1997) *The Open-Ended Approach: a new proposal for teaching mathematics*, Reston, Virginia, NCTM.

Belenky, M., Clinchy, B., Goldberger, N. and Tarule, J. (1986) *Women's Ways of Knowing: the development of self, voice and mind*, New York, Basic Books.

Bell, A. (1986) 'Diagnostic teaching: 2—Developing conflict-discussion lessons', *Mathematics Teaching*, vol. 116, p. 26–29.

Bell, A. (1987) 'Diagnostic teaching: 3—Provoking discussion', *Mathematics Teaching*, vol. 118, p. 21–23.

Bell, A., Brekke, G. and Swan, M. (1987a) 'Diagnostic teaching: 4—Graphical interpretation', *Mathematics Teaching*, vol. 119, pp. 56–59.

Bell, A., Brekke, G. and Swan, M. (1987b) 'Diagnostic teaching: 5—Graphical interpretation, teaching styles and their effects', *Mathematics Teaching*, vol. 120, pp. 50–57.

Bell, A. and Purdy, D. (1986) 'Diagnostic teaching', *Mathematics Teaching*, vol. 115, pp. 39–41.

Bennett, J. (1966) *The Dramatic Universe*, London, Routledge.

Berne, E. (1964) *Games People Play*, Harmondsworth, Penguin.

Biggs, J. and Collis, K. (1982) *Evaluating the Quality of Learning: the SOLO taxonomy*, New York, Academic Press.

Bills, L. and Rowland, T. (1999) 'Examples, generalisation and proof' in Brown, L. (ed.) *Making Meaning in Mathematics: visions of mathematics 2, advances in mathematics education*, no. 1, pp. 103–116, York, QED.

Boaler, J. (1997) *Experiencing School Mathematics: teaching styles, sex and setting*, Buckingham, Open University Press.

Boero, P., Garuti, R. and Mariotti, M. A. (1996) 'Some dynamic mental processes underlying producing and proving conjectures', *Proceedings of the 20th Conference of the International Group for the Psychology of Mathematics Education*, Valencia, vol. 2, pp. 121–128.

Brissenden, T. (1980) *Mathematics Teaching: theory and practice*, London, Harper and Row.

Brookes, W. (1966) *The Development of Mathematical Activity in Children: the place of the problem in this development*, Report prepared for the sub-committee on mathematical instruction of the British National Committee for Mathematics, Nelson, Lancs., ATM.

Brousseau, G. (1997) *Theory of Didactical Situations in Mathematics: didactiques des mathématiques, 1970–1990* in Balacheff, N., Cooper, M., Sutherland, R. and Warfield, V. (eds.), Dordrecht, Kluwer.

Brown, L. and Coles, A. (2000) 'Same/different: a 'natural' way of learning mathematics' in Nakahara, T. and Koyama, M. (eds) *Proceedings of the 24th Conference of the International Group for the Psychology of Mathematics Education*, Hiroshima, pp. 2–153.

Brown, S., Collins, A. and Duguid P. (1989) 'Situated cognition and the culture of learning', *Educational Researcher,* vol. 18, no. 1, pp. 32–41.

Brown, S. and Walter, M. (1983) *The Art of Problem Posing*, Philadelphia, Franklin Press.

Bruner, J. (1966) *Toward a Theory of Instruction*, Cambridge, Harvard University Press.

Bruner, J. (1986) *Actual Minds, Possible Worlds*, Cambridge, Harvard University Press.

Burger, W. and Shaunessy, J. (1986) 'Characterizing the van Hiele levels of development in geometry', *Journal for Research in Mathematics Education*, pp. 31–48.

Chevallard, Y. (1985) *La Transposition Didactique*, Grenoble, La Pensée Sauvage.

Christiansen, B. and Walther, G. (1986) 'Task and activity' in Christiansen, B., Howson, G. and Otte, M., *Perspectives in Mathematics Education*, Dordrecht, Reidel.

Claxton, G. (webref) http://www.early-education.org.uk/1newsletter000 699.htm

Cobb, P. (1988) 'The tension between theories of learning and instruction in mathematics education', *Educational Psychologist,* vol. 23, no. 2, pp. 87–103.

Cobb, P. (1994) 'Where is the mind? Constructivism and sociocultural perspectives on mathematical development', *Educational Researcher*, vol. 23, no. 7, pp. 13–20.

Cobb, P., Wood, T. and Yackel, E. (1990) 'Classrooms as learning environments for teachers and researchers' in Davis, R., Maher, C. and Noddings, N. (eds) *Constructivist Views on the Teaching and Learning of Mathematics*, Reston, Virginia, NCTM.

Cobb, P., Yackel, E. and Wood, T. (1992) 'Interaction and learning in mathematics situations', *Educational Studies in Mathematics*, no. 23, pp. 99–122.

Coles, A. and Brown, L. (1999) 'Meta-commenting: developing algebraic activity in a "community of inquirers"' in Bills, L. (ed.) *Proceedings of the British Society for Research into Learning Mathematics*, pp. 1–6, Warwick University, MERC.

Da Vinci, L. www.brainyquote.com/quotes/quotes/l/q140595.html

Davydov, V. (1990) *Types of Generalisation in Instruction* (trans. J. Teller) Reston, Virginia, NTCM.

Davis, R. (1966) 'Discovery in the teaching of mathematics' in Shulman, L. and Keislar, E. (eds) *Learning by Discovery: a reappraisal*, Chicago, Rand McNally.

Davis, B. (1996) *Teaching Mathematics: toward a sound alternative*, London, Garland.

Davis, B. (webref) *Mathematics Teaching: moving from telling to listening*, phenomenological research paper, Textorium, http://www.atl.ualberta.ca/po/main.cfm

Davis, R. Maher, C. and Noddings, N. (eds) (1990) *Constructivist Views on the Teaching of and Learning of Mathematics*, Reston, Virginia, NTCM.

Denvir, B. and Brown, M. (1986a) 'Understanding number concepts in low attaining 7–9-year-olds', *Educational Studies in Mathematics*, vol. 17, no. 1, pp. 15–36.

Denvir, B. and Brown, M. (1986b) 'Understanding number concepts in low attaining 7–9-year-olds' part II, *Educational Studies in Mathematics*, vol. 17, no. 2, pp. 143–164.

Dewey, J. (1902) *The Child and The Curriculum*, (reprinted 1971 as *The Child and The Curriculum and The School and Society*), Chicago, Chicago Press.

DfEE (1999a) *National Numeracy Strategy: framework for teaching mathematics from reception to year 6*, London, DfEE.

DfEE (1999b) *National Numeracy Strategy: mathematical vocabulary*, London, DfEE.

Dienes, Z. (1963) *An Experimental Study of Mathematics Learning*, London, Hutchinson.

Dubinsky, E. (1991) 'Reflective abstraction in mathematical thinking', in Tall, D. (ed.) *Advanced Mathematical Thinking*, Dordrecht, Kluwer.

Dweck, C. (1999) *Self-Theories: their role in motivation, personality and development*, Philadelphia, Psychology Press.

Dyrszlag, Z. (1984) 'Sposoby Kontroli Rozumienia Pojec Matematycznych', *Oswiata i Wychowanie 9*, B, pp. 42–43.

Festinger, L. (1957) *A Theory of Cognitive Dissonance*, Stanford, Stanford University Press.

Floyd, A., Burton, L., James, N. and Mason, J. (1981) *Developing Mathematical Thinking*, Milton Keynes, The Open University.

Frankenstein, M. (1989) *Relearning Mathematics: a different third R— radical math*, London, Free Association.

Freudenthal, H. (1983) *Didactical Phenomenology of Mathematical Structures*, Dordrecht, Reidel.

Freudenthal, H. (1991) *Revisiting Mathematics Education: China lectures*, Dordrecht, Kluwer.

Gadamer, H.-G. (1975) *Truth and Method*, New York, The Seabury Press.

Gadamer, H.-G. (1990) *Truth and Method*, New York, Continuum.

Gattegno, C. (1970) *What We Owe Children: the subordination of teaching to learning*, London, Routledge and Kegan Paul.

Gattegno, C. (1987) *The Science of Education, Part I: Theoretical considerations*, New York, Educational Solutions.

Gibson, J. (1977) 'The theory of affordances' in Shaw, R. E. and Bransford, J. (eds) *Perceiving, Acting, and Knowing*, Mahwah, Lawrence Erlbaum Associates.

Gibson, J. (1979) *The Ecological Approach to Visual Perception*, London, Houghton Mifflin.

Godfrey, C. and Siddons, A. W. (1931) *The Teaching of Elementary Mathematics*, Cambridge, Cambridge University Press.

Goodman, N. (1978) *Ways of World Making*, Boston and London, Harvester.

Gravemeijer, K. (1994) *Developing Realistic Mathematics Education*, Utrecht, Freudenthal Institute.

Gray, E. M. and Tall, D. O. (1994) 'Duality, ambiguity and flexibility: a proceptual view of simple arithmetic', *The Journal for Research in Mathematics Education*, vol. 26, no. 2, pp.115–141.

Hadas, N., Hershkowitz, R. and Schwarz, B. (2002) 'Analyses of activity design in geometry in the light of student actions', *Canadian Journal of Science, Mathematics and Technology Education*, vol. 2, no. 4, pp. 529–552.

Hamilton, E. and Cairns, H. (eds) (1961) (trans. Guthrie, W.) *Plato: the collected dialogues including the letters*, Bollingen Series LXXI, Princeton, Princeton University Press.

Hauser, M. (2001) *Wild Minds: what animals really think*, Harmondsworth, Penguin.

Healy, L. and Hoyles, C. (2000) 'Study of proof conceptions in algebra', *Journal for Research in Mathematics Education*, vol. 31, no. 4, pp. 396–428.

Hewitt, D. (1996) 'Mathematical fluency: the nature of practice and the role of subordination', *For the Learning of Mathematics*, vol. 16, no. 2, pp. 28–35.

Holt, J. (1964) *How Children Fail*, Harmondsworth, Penguin.

Hopkins, C. (1990) 'A conference tale', *Mathematics Teaching*, no. 132, pp. 20–21.

Houssart, J. (1999) 'Seeing the pattern and seeing the point' in Bills, L. (ed.) *Proceedings of the British Society for Research into Learning Mathematics*, pp. 73–78, Warwick University, MERC.

Jaworski, B. (1994) *Investigating Mathematics Teaching: a constructivist enquiry*, London, Falmer Press.

Johnson, D., Adhami, M. and Shayer, M. (1997) 'Does 'CAME' work? Summary report on Phase 2 of the Cognitive Acceleration in Mathematics Education (CAME) Project' in Morgan, C. (ed.) *Proceedings of the British Society for Research into Learning Mathematics*, pp. 26–31, and Bristol, Bristol University.

Kangshen, S., Crossley, J. and Lun, A. (1999) *The Nine Chapters on the Mathematical Art: companion and commentary*, Oxford, Oxford University Press.

Kilpatrick, J., Swafford, J. and Findell, B. (eds) (2001) *Adding It Up: helping children learn mathematics*, Washington, National Academy Press.

Krutetskii, V. (1976) (Teller, J., trans.) in Kilpatrick, J. and Wirszup, I. (eds) *The Psychology of Mathematical Abilities in School Children*, Chicago, University of Chicago Press.

Laborde, C. (1989) 'Audacity and reason: French research in mathematics education', *For the Learning of Mathematics*, vol. 9, no. 3, pp. 31–36.

Lakatos, I. (1976) *Proofs and Refutations*, Cambridge, Cambridge University Press.

Lave, J. (1988) *Cognition in Practice: mind, mathematics and culture in everyday life*, Cambridge, Cambridge University Press.

Lave, J. and Wenger, E. (1991) *Situated Learning: legitimate peripheral participation*, Cambridge, Cambridge University Press.

Legrand, M. (1993) *Débate Scientifique en Cour de Mathématiques*, Repères IREM, No. 10, Topiques Edition.

Leont'ev, A. (1981a) 'The problem of activity in psychology' in Wertsch, J. (ed.) *The Concept of Activity in Soviet Psychology*, Sharpe, Armonk.

Leont'ev, A. (1981b) *Psychology and the Language Learning Process*, Oxford, Pergamon.

Lerman, S. (1989) 'Constructivism, mathematics and mathematics education', *Educational Studies in Mathematics*, vol. 20, pp. 211–223.

Lerman, S. (1996) 'Intersubjectivity in mathematics learning: a challenge to the radical constructivist paradigm', *Journal for Research in Mathematics Education*, vol. 27, no. 2, pp. 133–150.

Li, S. (1999) 'Does practice make perfect?', *For The Learning of Mathematics*, vol. 19, no. 3, pp. 33–35.

Love, E. and Mason, J. (1992) *Teaching Mathematics: action and awareness*, Milton Keynes, The Open University.

Mcintosh, A. and Quadling, D. (1975) 'Arithmogons', *Mathematics Teaching*, no. 70, pp. 18–23.

Marton, F. and Booth, S. (1997) *Learning and Awareness*, Mahwah, Lawrence Erlbaum Associates.

Mason, J. (1979) 'Which medium, which message', *Visual Education*, Feb. 1979, pp. 29–33.

Mason, J. (2001) 'Mathematical teaching practices at tertiary level: Working Group report' in Holton, D. (ed.) *The Teaching and Learning of Mathematics at University Level: An ICMI Study*, Dordrecht, Kluwer.

Mason, J., Burton, L. and Stacey, K. (1982) *Thinking Mathematically*, London, Addison Wesley.

Mason, J. and Houssart, J. (2000) 'Arithmogons: a case study in locating the mathematics in tasks', *Primary Teaching Studies*, vol. 11, no. 2, pp. 34–42.

Mathematical Association (1987) *Maths Talk*, Cheltenham, Stanley Thornes.

Maturana, H. and Varela, F. (1988) *The Tree of Knowledge: the biological roots of human understanding*, Boston, Shambala.

Mehan, H. (1986) '"What time is it Denise?": asking information questions in classroom discourse' in Hammersley, M. (ed.) *Case Studies in Classroom Research*, Buckingham, Open University Press, pp. 85–103.

Mellin-Olsen, S. (1987) *The Politics of Mathematics Education*, Dordrecht, Reidel.

Movshovits-Hadar, N. (1988) 'Surprise', *For the Learning of Mathematics*, vol. 8, no. 3, pp. 34–40.

Nunes, T., Schliemann, A. D. and Carraher, D.W. (1993) *Street Mathematics and School Mathematics*, New York, Cambridge University Press.

Piaget, J. (1950) *The Psychology of Intelligence*, London, Routledge and Kegan Paul.

Piaget, J. (1971) *Biology and Knowledge*, Chicago, University of Chicago Press.

Piaget, J. (1972) (trans. Mays, W.) *Principles of Genetic Epistemology*, London, Routledge and Kegan Paul.

Piaget, J. (1977) (Rosen, A., trans.) *The Development of Thought: equilibration of cognitive structures*, New York, Viking.

Pirie, S. and Kieren, T. (1989) 'A recursive theory of mathematical understanding', *For the Learning of Mathematics*, vol. 9, no. 4, pp. 7–11.

Pirie, S. and Kieren, T. (1994) 'Growth in mathematical understanding: how can we characterise it and how can we represent it?', *Educational Studies in Mathematics*, vol. 26, no. 2–3, pp. 165–190.

Pirie, S. E. B. and Schwarzenberger, R. L. E. (1988) 'Mathematical discussion and mathematical understanding', *Educational Studies in Mathematics*, vol. 4, no. 19, pp. 459–470.

Prestage, S. and Perks, P. (1992) 'Making choices (part 2): "... not if you're a bear"', *Mathematics in Schools*, vol. 21, no. 4, pp. 10–11.

Raymond, L. (1972) *To Live Within*, London, George Allen and Unwin.

Sàenz-Ludlow, A. and Walgamuth, C. (2001) 'Question- and diagram-mediated mathematical activity: a case in a fourth grade classroom', *Focus on Learning Problems in Mathematics*, vol. 23, no. 4, pp. 27–40.

Schmidt, W. H. (ed.) (1996) *Characterizing Pedagogical Flow: an investigation of mathematics and science teaching in six countries*, Dordrecht, Kluwer.

Sfard, A. (1994) 'Reification as the birth of metaphor', *For the Learning of Mathematics*, vol. 14, no. 1, pp. 44–55.

Sfard, A., Nesher, P., Streefland, L., Cobb, P. and Mason, J. (1998) 'Learning mathematics through conversation: is it as good as they say?', *For the Learning of Mathematics*, vol. 18, no. 1, pp. 41–51.

Sierpinska, A. (1994) *Understanding in Mathematics*, London, Falmer Press.

Skemp, R. (1976) 'Relational understanding and instrumental understanding', *Mathematics Teaching*, no. 77 (December), pp. 20–26.

Skemp, R. (1979) *Intelligence, Learning and Action*, Chichester, Wiley.

Skvovemose, O. (1994) *Towards a Philosophy of Critical Mathematics Education*, Dordrecht, Kluwer.

Snyder, B. (1970) *The Hidden Curriculum*, New York, Alfred-Knopff.

Spencer, H. (1929) *Education: intellectual, moral, and physical*, Thinker's Library No. 2, London, Watts.

Steffe, L., von Glasersfeld, E., Richards, J. and Cobb, P. (1983) *Children's Counting Types: philosophy, theory, and application*, New York, Praeger Scientific.

Stein, S. (1987) 'Gresham's law: algorithm drives out thought', *For the Learning of Mathematics*, vol. 7, no. 2, pp. 2–4.

Stigler, J. and Hiebert, J. (1999) *The Teaching Gap: best ideas from the world's teachers for improving education in the classroom*, New York, Free Press.

Tahta, D. (1972) *A Boolean Anthology: selected writings of Mary Boole on mathematics education*, Derby, Association of Teachers of Mathematics.

Tahta, D. (1981) 'Some thoughts arising from the new Nicolet films', *Mathematics Teaching*, no. 94, pp. 25–29.

Tahta, D. (1991) 'Understanding and Desire' in Pimm, D. and Love, E. (eds) *Teaching and Learning School Mathematics*, pp. 220–246, London, Hodder and Stoughton.

Tall, D. and Vinner, S. (1981) 'Concept image and concept definition in mathematics with particular reference to limits and continuity', *Educational Studies in Mathematics*, vol. 12, no. 2, pp. 151–169.

The Open University (1980) *PME233: Real Problem Solving*, Milton Keynes, The Open University,

Thurston, W. (1994) 'Proof and progress in mathematics', *Bulletin of the American Mathematical Society*, vol. 30, no. 7, pp. 161–177; reprinted in *For the Learning of Mathematics*, vol. 15, no. 1, 1995, pp. 29–37.

TIMSS (1997) *Mathematics Achievement in the Primary School Years*. Boston College, Chestnut Hill, MA.

van Hiele, P. (1986) *Structure and Insight: a theory of mathematics education*, Orlando, Academic Press.

Vergnaud, G. (1983) 'Multiplicative structures' in Lesh, R. and Landau, M. (eds) *Acquisition of Mathematics Concepts and Structures*, pp. 127–174, New York, Academic Press.

von Glasersfeld, E. (1995) *Radical Constructivism: a way of knowing and learning*, London, Falmer Press.

Vygotsky, L. (1934/1986) *Thought and Language*, Cambridge, MIT Press.

Vygotsky, L. (1978) *Mind in Society: the development of the higher psychological processes*, London, Harvard University Press.

Walkerdine, V. (1988) *The Mastery of Reason*, London, Routledge and Kegan Paul.

Watson, A. (2000) 'Going across the grain: mathematical generalisation in a group of low attainers', *Nordisk Matematikk Didaktikk (Nordic Studies in Mathematics Education)*, vol. 8, no. 1, pp. 7–22.

Watson, A. (2002) 'Instances of mathematical thinking among low attaining students in an ordinary secondary classroom', *Journal of Mathematical Behaviour*, vol. 4, no. 20, pp. 461–475.

Watson, A. and Mason, J. (1998) *Questions and Prompts for Mathematical Thinking*, Derby, Association of Teachers of Mathematics.

Watson, A. and Mason, J. (2002) 'Student-generated examples in the learning of mathematics', *Canadian Journal of Science, Mathematics and Technology Education*, vol. 2, no. 2, pp. 237–249.

Wertsch, J. (ed.) (1981) *The Concept of Activity in Soviet Psychology*, Sharpe, Armonk.

Whitehead, A. (1919) (12th impression, reprinted 1948) *An Introduction to Mathematics*, London, Oxford University Press.

Wood, P., Bruner, J. and Ross, G. (1976) 'The role of tutoring in problem solving', *Journal of Child Psychology and Psychiatry*, vol. 2, no. 17, pp. 89–100.

INDEX

MORE ABOUT TARQUIN

Tarquin provides a wide variety of innovative mathematics materials including:

- Books
- Posters
- Software
- DIME
- Zome
- And lots more…

Tarquin Publications
99 Hatfield Road
St Albans
AL1 4JL
UK

Tel +44 01727 833866
Fax +44 0845 456 6385
Email: info@tarquinbooks.com

www.tarquinbooks.com

Lightning Source UK Ltd.
Milton Keynes UK
UKOW020349220212

187722UK00001B/12/P